A WARMHEARTED WELCOME TO
"PERSONAL CARE FOR PEOPLE
WHO CARE"

♥

You're looking at a very special heart. It stands for an end to pain and suffering for millions of animals who will no longer be sacrificed for the sake of vanity. It represents the combined commitment and effort of many people– those who care about animals, innovative scientists who have helped to develop alternative methods for testing products, and progressive corporate leaders who now develop and sell products without the hidden ingredient of cruelty. It also stands for informed consumers who understand that their choices make a difference. This heart stands for new hope, caring and compassion. This heart stands for cruelty-free.™

We invite you to learn about the hundreds of companies and their fine products which are not tested on animals. You'll find them listed in this book. Unfortunately, hundreds more still test on animals–and millions of animals suffer needlessly because of it. You can make a difference. By purchasing only cruelty-free products you can convince companies that test on animals to have a change of heart.

On behalf of all of us at the National Anti-Vivisection Society, we thank you for caring enough to do something for the animals.

Have a heart. Buy cruelty-free.

Peggy Cunniff

Mary Margaret Cunniff
Executive Director

THE NATIONAL ANTI-VIVISECTION SOCIETY MISSION STATEMENT

The National Anti-Vivisection Society is dedicated to abolishing the exploitation of animals used in research, education and product testing.

NAVS promotes greater compassion, respect and justice for animals through educational programs based on respected ethical and scientific theory and supported by extensive documentation of the cruelty and waste of vivisection. NAVS' educational programs are directed at increasing public awareness about vivisection, identifying humane solutions to human problems, developing alternatives to the use of animals and working with like-minded individuals and groups to effect changes which help to end the suffering inflicted on innocent animals.

The information provided in this book is based on the most current data available at the time of printing. Companies reserving the right to conduct animal tests at any time in the future have been listed as companies that test. For the most accurate information, contact the company directly.

The National Anti-Vivisection Society is a national, not-for-profit organization.

Cover Photo: Lyal Lauth, 456 N. Leavitt, Chicago, IL 60612. Phone: 312-829-9800.
Cover Design: Frothingham Communications, A Full-Service Creative Services
    Company, 415 Washington Ave., Wilmette, IL 60091. Phone: 708-251-2073.
Cosmetics courtesy of Elaine Gayle from Juvenesse, Chicago, IL.
Hand woven makeup bag created by Barbara Brooks, Pittsburgh, PA.
Photo model: Kathryn Frothingham, age 8.

ISBN 1-888635-00-2 ♻ Please recycle this book.

TABLE OF CONTENTS

PERSONAL CARE FOR PEOPLE WHO CARE:
BUYING FROM COMPANIES THAT HAVE A HEART

The National Anti-Vivisection Society (NAVS) is proud to introduce the eighth edition of its guide to being a compassionate consumer. *Personal Care for People Who Care* contains the most current information available on the product testing policies of companies that manufacture and distribute cosmetic, personal care, household, and companion animal products. We hope this book will help you find–and use–cruelty-free™ products.

A wide range of cruelty-free products exists. You'll find cruelty-free products in local department stores, pharmacies and retail outlets. If they are not available, please encourage your store manager to carry cruelty-free merchandise. You may also wish to contact manufacturers, request their catalogs, and order from them directly. If manufacturers don't sell directly to consumers, they may refer you to stores in your area that carry their products. If you request product information from a manufacturer, please direct your call to the consumer affairs department.

Cruelty-free mail order companies give you the opportunity to purchase the products of many manufacturers from a single source. Retail stores and health food stores also sell cruelty-free items, but not necessarily exclusively. Be sure to take this book with you when you visit your local retail and health food stores.

Reading *"Personal Care"* will help you find answers to the most frequently asked questions regarding cruelty-free products and what you can do to help stop the use of animals in product testing.

For more information on NAVS and its programs, please contact us at:

> The National Anti-Vivisection Society
> 53 West Jackson Boulevard, Suite 1552
> Chicago, IL 60604
> (312) 427-6065,
> (800) 888-NAVS (6287)
> Fax: (312) 427-6524
> E-mail: navs@navs.org

Also look for us on the Internet at our World Wide Web site:

> http://www.navs.org

A HEART-TO-HEART QUESTION AND ANSWER SESSION WITH NAVS

Q: WHO IS NAVS?
A: The National Anti-Vivisection Society (NAVS) began in 1929 as a small, impassioned group of individuals dedicated to ending animal exploitation. Since then, NAVS has been at the forefront of the animal advocacy movement and has grown in membership and purpose. Today it is a vital, non-profit educational organization, committed to promoting public awareness of the cruelty and waste of animal experimentation, as well as advancing more humane and better methods of scientific investigation. We aim to end the needless suffering of animals through positive and productive programs that help educate the public, support passage of animal protection legislation, and fund alternatives to animal research.

Q: WHAT DOES "VIVISECTION" MEAN?
A: In simplest terms, vivisection is the act of cutting into, dissecting or harming the body of a living animal, especially for the purpose of scientific research. The National Anti-Vivisection Society is opposed to this practice, and works to abolish the use of animals in the areas of product testing, education and research. Look to the following pages to learn about the various programs that NAVS offers to help eliminate animal suffering.

Q: ISN'T MOST ANIMAL TESTING DONE TO BENEFIT MEDICAL RESEARCH?
A: Absolutely not. It has been estimated that biomedical research accounts for only 27 percent of the animals currently being used in the name of science. The majority of this research isn't even published, which means it has little or no value. As a result, animal experimentation has had little significant influence on the health of the nation. The remaining 73 percent of animal experimentation falls into the categories of product testing and education. An estimated 16 million animals are used each year in manufacturers' labs and in classrooms. In other words, at least 16 million animals could be saved each year without affecting biomedical research.

Q: AREN'T TESTS ON ANIMALS NECESSARY TO ENSURE PRODUCT SAFETY?

A: No. Testing products and their ingredients on animals is no guarantee of product safety. Animal testing merely determines the level of toxicity for that particular nonhuman animal. In fact, toxic products that have been tested on animals are regularly introduced into the marketplace, and are only required to carry a warning label. That's because no amount of animal testing can change the fact that many of these products are harmful if ingested or used in a way not intended by the manufacturer. In other words, many of these products that have been animal-tested are no less deadly if a child eats or drinks them accidentally.

Tests on nonhuman animals, for instance, cannot accurately predict an allergic reaction by some humans. Products found to be safe on nonhuman animals can even cause minor to serious side effects on people.

Q: DOESN'T THE GOVERNMENT REQUIRE ANIMAL TESTING FOR COSMETIC AND HOUSEHOLD PRODUCTS?

A: Another myth. The Food and Drug Administration strongly urges cosmetic and household product manufacturers to conduct tests that substantiate the safety of their products. However, the FDA actually encourages the use of non-whole animal techniques.

Q: WHAT OTHER FORMS OF TESTING ARE AVAILABLE?

A: There are many effective alternatives to animal testing, from tissue cultures to computer and mathematical models. And hundreds of companies throughout the country are currently using these cruelty- free tests successfully. For more details, see the section entitled, "Product Testing: Why It's Time For A Change of Heart."

Q: HOW IS "CRUELTY-FREE" DEFINED BY NAVS?

A: Cruelty free refers to the product testing practices of a company. For a company to be recognized as cruelty-free, it must discontinue animal testing of any form and have no corporate intention of resuming it. In addition, the ingredients or raw materials acquired from outside sources must not be subjected to animal testing. Although our definition of cruelty-free does not mean "vegan" and therefore does not take into consideration animal by-products or ingredients that may be contained in a company's products, those companies whose products contain no animal ingredients are identified in this book by a ☆.

Q: WHAT IS NAVS DOING?

A: By working with the public, scientific and legislative communities, we're able to stay abreast of technological advances and new information relating to biomedical research, product testing and education. That means we're able to use up-to-date statistical and technical data to substantiate our position against animal experimentation—and make a positive impact. Here are just a few of the programs and activities NAVS sponsors and supports.

The NAVS Bulletin. This quarterly publication keeps members informed of NAVS activities and relevant issues across the country.

Target: NOAH. Exclusively available to NAVS members, **Neighbors Organized for Animals and Health** provides local groups and individuals with a manual that can identify what kind of animal experimentation takes place in their communities and provides strategies to, and suggestions for, things they can do to help end the cruelty to animals.

The International Foundation for Ethical Research (IFER). Conceived and supported by NAVS, this foundation provides grants to scientists for research projects that offer promising, viable alternatives to animal research. Animal advocates and innovative scientists are working together in very constructive ways. For example, a multiple assay test funded by IFER is currently being used by several pharmaceutical and cosmetic companies.

Resources. In conjunction with The John Marshall Law School, NAVS has founded the first National Research Library for Animal Advocacy in Chicago, Illinois.

Alternatives to Dissection. The Society is the sole sponsor of the NAVS Dissection Hotline (1-800-922-FROG) a toll-free service for students objecting to dissection. The Hotline provides information, support and counseling to thousands of students and teachers who believe life science instruction is better achieved without sacrificing animals. In addition, NAVS offers an alternative to dissection with our free loan of anatomically accurate three-dimensional frog and fetal pig models to teachers and students nationwide. Alternatives like this can spare the lives of thousands of animals each year. We hope these efforts will eventually foster a more just and humane educational system.

Informational Videos. Educational videos are available from NAVS on a two-week loan basis, including the NAVS-produced "Advances in Humane Education: Alternatives in Biology." Call or write for our resource guide.

Public Awareness. NAVS has produced award-winning public service announcements for television, radio and print-media that have helped to inform millions of Americans about animal research, product testing and dissection issues. Dedicated NAVS staff members make credible and compelling presentations to schools and universities across the country, TV, radio, and letters to the editor, and to the general public as well. NAVS believes it is imperative to provide members of the public with a better understanding of vivisection, and the humane cost-saving alternatives to animal research. Please contact us if you would like a NAVS staff member to speak at a meeting or classroom presentation.

Another initiative has been an annual *Art for the Animals Classic,* which has provided a creative way for many to express and share their concern for animals.

Q. WHAT CAN I DO TO HELP?
A. First, become a more informed animal advocate and consumer. Find out all you can about the cruelty and waste of vivisection—and share that information with your family, friends and co-workers. We would also ask you to give thoughtful consideration to becoming a member of NAVS.

Since NAVS is a non-profit organization, we can't survive without your help. Funding is a vital means of supporting and continuing our public awareness programs. While these educational programs are making a real difference in sparing animals from pain and suffering, much more needs to be done. The collective efforts of caring and informed individuals play a critical role in expanding our efforts. But funding is only one way to lend your support. See the following section for other ways you can help out.

GETTING TO THE HEART OF THE MATTER: WHAT YOU CAN DO TO MAKE A DIFFERENCE

This book, *Personal Care For People Who Care,* is considered the best guide available on the subject of cruelty-free cosmetic, personal care, household, and companion animal products. Over the past year, we've been able to welcome more than 150 new cruelty-free companies to our listing.

LETTERS WITH HEART

Readers of previous editions of this book have played an important role by contacting companies that have not answered NAVS inquiries. Please help us persuade the companies listed under "Did Not Respond," and urge them to complete the NAVS' surveys. You may want to consider the following questions as guidelines for your letter.

• Does your company or any outside source test your products or their ingredients on animals?

• How long (if applicable) have you not tested products or their ingredients on animals, e.g., past six months, past year, past five years, never?

• Do any of your products contain animal ingredients (e.g. lanolin, honey, bees' wax)? Do any of your products contain slaughterhouse products (e.g. collagen, gelatin)?

• Is your company a subsidiary, division or parent company of another company? If so, please identify this company.

• Please list some alternatives your company is using in place of animal testing.

Of course a heart-felt message in your own words is still the best way to reach a company.

JOIN NAVS AND ENCOURAGE OTHERS TO DO THE SAME

Become a member of the National Anti-Vivisection Society to help us fight vivisection. Use the application form in the back of the book or contact NAVS for more information.

USE AND DISTRIBUTE THIS BOOK AT EVERY OPPORTUNITY

By helping us distribute "Personal Care" and encouraging store owners to carry it, you are helping more people make informed choices by educating them about the product testing practices currently in use by manufacturers. Please consider ordering additional copies and donate them to your local library and schools. Help us be heard. Spread the word.

USE YOUR ECONOMIC POWER ETHICALLY

Companies respond to public demand. Buy products only from companies who DO NOT subscribe to animal testing. And don't invest in the stock of companies who test their products on animals. If you already hold investments in any of these companies, use your power as a stockholder to encourage a change in policy.

APPEAL TO LEGISLATORS

Inform your elected representatives how you feel about the use of live animals for product testing. Encourage them to support bills to end these practices.

NAVS is committed to promoting a cruelty-free society, but we can't do it alone. Millions of animals are dying for your help. Have a heart. Lend a hand. Make a difference.

FEDERAL FACTS, ACTS, AND DEFINITIONS YOU SHOULD TAKE TO HEART

The products listed in this book are divided into four primary groups: cosmetics, personal care items, household products, and companion animal products. The following histories and definitions are included to provide you with a clearer understanding of federal regulations and manufacturer responsibilities.

COSMETICS
U.S. Food and Drug Administration (FDA)

In 1938 the Federal Food, Drug and Cosmetic Act defined cosmetics as "articles intended to be applied to the human body for cleansing, beautifying, promoting attractiveness, or altering the appearance without affecting the body's structure or functions." Included under this definition are perfumes, lotions, eye and facial make-up preparations, skin creams, lipsticks, fingernail polishes, shampoos, permanent waves, hair color, toothpastes, deodorants, and any ingredient used as a component of a cosmetic product.

Cosmetics marketed in the United States, whether made here or imported, must comply with the Fair Packaging and Labeling Act, the Federal Food, Drug and Cosmetic Act, and the regulations issued under the authority of these laws.

The Federal Food, Drug and Cosmetic Act does not require that cosmetic manufacturers or marketers test their products for safety. But the FDA strongly urges cosmetic manufacturers to conduct whatever toxicological tests are appropriate to substantiate the safety of their cosmetics. If the safety of a cosmetic is not adequately substantiated, the product may be considered misbranded and may be subject to regulatory action unless the label bears the following statement: "Warning - The safety of this product has not been determined (21 CFR 740.10)."*

With the exception of color additives and a few prohibited ingredients, a cosmetic manufacturer may use essentially any raw material as a cosmetic ingredient and market the product without approval.

COSMETICS THAT ARE CONSIDERED DRUGS

Cosmetics and personal care products that are also intended to treat or prevent disease, or affect the structure or functions of the human body, are considered drugs. These products must comply with the drug requirements of the FDA. Animals almost always are utilized as the test models. For example, if a fluoride toothpaste claims to make teeth whiter, it is considered a drug. Suntan preparations intended to protect against sunburn, deodorants that are also antiperspirants and anti-dandruff shampoos fall into this category.

---

* Section 21 of the Code of Federal Regulations, part 740.10

CONTACT LENS PRODUCTS

According to information obtained from the FDA, contact lenses and lens care products are medical devices, not pharmaceuticals. They differ from pharmaceuticals in that they achieve their primary intended purpose by a physical means, rather than by chemical actions, as is the case with pharmaceuticals.

The FDA has not published any regulations that require animal testing of contact lenses or lens care products. However, the FDA has made available to manufacturers testing guidance documents. These documents do not carry the weight of regulations, but they do recommend preclinical testing using animals. When alternative procedures are chosen by a manufacturer, justification must be provided which clearly explains why an alternate test procedure is an acceptable substitute which has been validated and accepted by the scientific community.

For further information on the FDA, contact them at:

Food and Drug Administration
Consumer Inquiries Staff
HFE-88, Room 16-63
5600 Fishers Lane
Rockville, MD 20857
(301) 443-3170

HOUSEHOLD PRODUCTS

U.S. Environmental Protection Agency (EPA)

The United States Environmental Protection Agency requires animal testing in some instances, as in the case of household insecticides and other products.

Consumer Product Safety Commission

In 1972, Congress passed the Consumer Product Safety Act. This in turn activated the Consumer Product Safety Commission the following year. The Commission was directed to "protect the public against unreasonable risks of injuries and deaths associated with consumer products."

Under the Consumer Product Safety Act, Congress granted the Commission broad authority to issue and enforce safety standards prescribing performance requirements, warnings or instructions for use of consumer products. Household products are labeled according to the ingredients in the product. Labeling of a household product comes under the jurisdiction of one or both of the following acts: The Federal Hazardous Substances Act and the Poison Prevention Packaging Act of 1970. Household products include such items as dishwashing liquids, furniture polishes, floor and oven cleaners and laundry detergents.

For further information concerning the Consumer Product Safety Commission please contact the office nearest you.

Central Region Center/CPSC
230 South Dearborn Street, Rm. 2945
Chicago, IL 60604
(312) 353-8260

Eastern Regional Center/CPSC
6 World Trade Center
Vesey Street, 3rd Floor
New York, NY 10048
(212) 264-1125

Western Regional Center/CPSC
600 Harrison, Rm. 245
San Francisco, CA 94017
(415) 744-2966

PRODUCT TESTING: WHY IT'S TIME FOR A CHANGE OF HEART

Draize Eye & Skin Irritancy Tests. Heartless and senseless.

The Draize Test was introduced more than 45 years ago by Food and Drug Administration toxicologist John H. Draize. It is used to measure the harmfulness of chemicals found in household products and cosmetics by observing the damage they cause to the eyes and skin of animals. Rabbits are often used because they are inexpensive, easy to handle, and have large eyes that make any irritation easy to observe. The animals are restrained, administered the chemicals, and the damage is observed and recorded over the course of three to 21 days. The test compounds often cause irreparable damage to the rabbits' eyes, leaving them ulcerated and bleeding. All of the animals are killed at the end of the test period.

For skin-irritancy tests, animals are immobilized while test substances are applied to shaved and abraded skin. Skin is abraded by firmly and repeatedly pressing adhesive tape to the animal's body and then quickly stripping it off.

These tests, and others like them, are used only to measure the harmfulness of products and their ingredients. They do not provide information for effective treatment of injuries that may result from product use. Many companies market products that have been shown to be irritants in Draize tests, and merely provide warning labels on the package. Other companies have never used Draize tests and have no difficulty marketing their products. "It appears that Draize testing is part of the process by which manufacturers avoid legal liability for damages caused by their products" (Physicians Committee for Responsible Medicine, 1988).

Lethal Dose 50 Percent (LD-50). Cold-hearted, outdated and unnecessary.

In the 1920s, the classic Lethal Dose 50 percent test was developed to measure the acute toxicity of certain ingredients by their effect on live animals. This test is used to determine the dosage required to kill 50 percent of a population of test animals, usually 200 or more, within a specified time frame. Every year, 4 to 5 million animals are force-fed, injected with, forced to inhale, or otherwise exposed to body lotions, drain cleaners, toothpastes, fabric bleaches and other toxic substances. As the concentration of a test substance increases, these animals may suffer convulsions, chronic diarrhea, massive bleeding and painful deaths.

The LD-50 test results vary from species to species, and are affected by a range of factors from age, sex, weight and diet, to humidity and temperature. Nausea, headache, dizziness, amnesia, depression and other sublethal toxic effects go undetected in nonhuman animal studies because animals are genetically programmed to avoid displays of physical weakness that might attract predators. And, of course, they cannot speak.

The LD-50 test does not provide information for the effective treatment of accidental poisoning. The test results have little bearing on consumer protection because these tests are not intended to keep toxic products off the market. Hundreds of thousands of accidental poisonings involving household and personal care products are reported to poison control centers every year. Emergency room physicians, however, don't use the results of LD-50 tests in their treatments.

ALTERNATIVES WITH HEART
Not only are nonhuman animals not appropriate test models because of their physiological differences from people, but safe, reliable, and cost-effective non-animal alternative test methods do exist, with the added advantage of reproducibility. In vitro (test tube) technology already provides an accurate alternative to Draize tests without the use of vertebrate animals.

The following is a brief description of some of the most commonly used non-animal product safety tests:

Agarose Diffusion Method—This method was originally designed for use over 25 years ago to determine the toxicity of plastics and other synthetic materials used in medical devices such as heart valves, intravenous lines, and artificial joints. In this test, human cells and a small amount of test material are placed in a container, separated by a thin layer of agarose, a derivative of the sea plant, agar. If the test material is an irritant, a zone of killed cells appears around the substance.

CAM (Chorioallantoic Membrane) Test—This test uses fertilized chicken eggs to evaluate eye irritancy by observing the reaction of the chorioallantoic membrane to test substances. Since the chorioallantoic membrane contains no nerve fibers, this test causes no discomfort or pain.

Computer and Mathematical Models—The irritancy of test substances can be predicted on the basis of physical and chemical structures and properties.

Eyetex—This test uses a protein alteration system to assess irritancy. A vegetable protein derived from the Jack bean mimics the cornea's reaction when exposed to foreign matter. The greater the irritation, the more opaque the solution becomes. Both the Eyetex and Skintex systems can be used to measure the toxicity of over 5,000 different materials.

Skintex—The Skintex formula is made from the yellowish meat of a pumpkin rind. Skintex imitates the reactions of human skin to foreign matter, measuring the irritancy of the material placed on the "skin".

Neutral Red Bioassay—In this test, neutral red, a water-soluble dye, is added to normal human skin cells in a 96-well tissue culture plate. A computer measurement of the level of "uptake" of the dye by the cells is used to indicate relative toxicity, eliminating observer bias.

Epipack Test—This test uses sheets of cloned human skin cells that accurately measure a human's reaction to skin irritation.

Non-animal alternatives can provide the needed data in a fraction of the time and cost of traditional animal tests. Cell and tissue cultures, mathematical models, and standardized biochemical assays are all currently being used effectively. Computer programs that predict product toxicity can provide even better results. The TOPKAT programs, developed by Health Designs Inc. (HDI), can predict which substances will cause skin and eye irritancy, or even more severe consequences. The FDA, the EPA and many private companies use the TOPKAT programs. More companies should be encouraged to adopt similar methods.

Scientists have concluded that no single alternative test available today can replace the Draize test, but they can rely instead on a battery of alternative tests. Companies that have chosen non-animal product testing methods generally do not rely solely on in vitro tests. They're incorporating these methods along with information on historical use and chemical structure, as well as human clinical studies.

These human studies include RIPT (Repeated Insult Patch Test), Maximization Test, Sun Protection Factor Assay, Clinical Safety Evaluation, Eye Area Products, and Clinical Safety Evaluations of Eye Area and Facial Products.

Of course, there are thousands of ingredients already known to be safe, and are on the U.S. GRAS (Generally Regarded As Safe) List that companies can use without testing.

HOW TO USE THIS BOOK

The purpose of this book is to guide you in purchasing cruelty-free products. All of the information is supplied to us directly from each company. If a product or company is owned by a major manufacturer, there will be a reference to "see parent company". The manufacturers and brands are listed in alphabetical order.

Frequently, manufacturers will purchase ingredients for their products from outside sources. With or without the manufacturer's knowledge, these ingredients may be subject to animal testing. As a result, we thought it only fair to distinguish "cruelty-free" companies from those unsure of ingredient and supplier testing status.

Personal Care is divided into two main sections:

1. Companies that responded.

2. Companies that did not respond.

KEY GUIDE:

❤ Companies that are CRUELTY-FREE—who DO NOT test their final products or ingredients on animals, nor contract out for such testing. In addition, ingredients or raw materials acquired from outside sources must not be subject to animal testing.

♡ Companies that do not test but whose ingredients MAY be tested on animals by their suppliers.

▼ Companies that DO test their final products or ingredients on animals.

☆ Products contain no animal ingredients

■ Manufacturer

● Distributor

✉ Mail order companies

For specific products manufactured and/or sold by cruelty-free companies, see the product reference guide later in this book.

Animal-Derived Ingredients:

NAVS primary criterion for cruelty-free status is that the company neither manufactures nor sells products that are tested on animals. However, some of the cruelty-free companies manufacture products containing animal-derived ingredients. If no products manufactured by or sold through the company contains any animal-derived ingredients, this symbol will appear after the company name: ☆

To obtain a list of ingredients that are derived from animal by-products, call us at 1-800-888-NAVS.

Moratorium Status:

Please note that there are manufacturers (Dial Corporation, Mary Kay Cosmetics) who are currently operating under a moratorium. As it pertains to this guide, a moratorium is defined as a suspension of animal testing until a permanent decision has been made concerning testing policies. Moratorium status will appear after the company product categories to denote this policy.

INDEX TO RESPONDING COMPANIES & THEIR ANIMAL TESTING POLICIES

♥ A'belir, Inc. ☆ ■ ●
305 Kingston Avenue
Daytona Beach, FL 32115
(800)771-7553,
(904)254-7553
Personal care

♥ A-1 Bleach
(see Austin's)

▼ A-200 Pyrinate
(see SmithKline Beecham)

♥ A-Retinol
(see C.E. Jamieson &
Co.,Ltd.)

▼ Aapri
(see Gillette)

♥ ABBA Products, Inc. ☆ ■
2010 Main Street, #1000
Irvine, CA 92714
(800)848-4475,
(714)851-3955
Personal care

♥ Abbaco, Inc. ■
230 Fifth Avenue
New York, NY 10001
(212)679-4550
Personal care

♥ Abby's Song of Life
Comfrey Goldenseal Salve
(see Song of Life)

♥ ABEnterprises ☆ ✉
145 Cortlandt St
Staten Island, NY 10302-
2048
(718)448-1526
Personal care, Household

♥ Abercrombie & Fitch
(see Gryphon Development)

♥ Abkit, Inc. ■
130 E.93rd St, Ste 1B
New York, NY 10128
(800)CAMOCARE,
(212)860-8358
Personal care, Household

♥ CRUELTY FREE
Does not test products
or ingredients on animals

♥ Ingredients MAY
be tested on
animals

▼ Tests products or
ingredients
on animals

❤ Abracadabra Bathcare ☆ ■
P.O. Box 1040
Guerneville, CA 95446
(707)869-0761,
In CA (800)745-0761
Personal care

❤ Acca Kappa
(see Goodebodies USA)

▼ Act Flouride Rinse
(see Johnson & Johnson)

❤ Actibath
(see Andrew Jergens)

❤ Action Labs ☆ ■ ●
2851 Voa Martens
Anaheim, CA 92806
(714)630-5941
Personal care

▼ Active Protection Make-up
(see Max Factor)

▼ Acuvue
(see Johnson & Johnson)

❤ Adios Polish Remover
(see Gena Labs)

❤ Adios Warm O Lotion
(see Gena Labs)

▼ Adorn
(see Gillette)

❤ Adrien Arpel ■
720 Fifth Avenue
New York, NY 10019
(212)333-7700
Cosmetics

❤ Advanced Research
Labs. ☆ ■
151 Kalmus Drive
Suite H-3
Costa Mesa, CA 92626
(800)966-6960
Personal care

❤ Advantage Wonder
Cleaner ■ ●
16615 S. Halsted St.
Harvey, IL. 60426
(800)323-6444,
(708)333-7644
Household

❤ ADWE Laboratories ● ■
141 20th Street
Brooklyn,NY 11232
(718)788-6838
*Cosmetics, Personal care,
Household*

▼ Aerowax
(see Reckitt & Colman)

☆ Products contain no
animal or animal
derived ingredients
　　　■ Manufacturer　● Distributor　⊠ Mail Order

♥ Affection Company ☆
1408 Brooks Road
Grass Valley,CA 95945
Personal care

♥ AFM Enterprises ☆ ■
350 West Ash Street, #700
San Diego, CA 92101
(714)781-6860
Personal care, *Household*

▼ Afta
(see Mennen)

▼ Age Shield Moisture
Treatment for Hands
(see Max Factor)

♥ Agree
(see Dep Corporation)

♡ AHC Pharmacal, Inc. ☆ ■
888 West 16th Street
Newport Beach, CA 92669
(714)631-0149
Cosmetics, *Personal care*

▼ Ahimsa Natural
Care Ltd. ☆ ■
1110 Sheppard Ave., East
North York, Ontario
Canada M2K 2W2
(416)667-1363
Personal care

♥ Aida Thibiant
(see Francosmetics Int'l)

▼ Aim
(see Chesebrough-Pond's)

♥ Air Theraphy
(see Mia Rose Products)

♡ Aiyana Skin Care System
(see Oxyfresh USA)

♥ AJ Funk & Company ☆ ■
1471 Timber Drive
Elgin, IL 60123
(708)741-6760
Household

▼ Ajax
(see Colgate-Palmolive)

♥ AKA Saunders, Inc. ■ ●
1200 Fifth Street
Berkeley, CA 94710
(510)528-0162
Personal care

♥ Al Stephans
(see La Dove)

♥ Alba Naturals, Inc. ■
P.O. Box 40339
Santa Barbara, CA
93140-0339
(800)347-5211
Personal care

▼ Alberto-Culver Company
2525 Armitage Avenue
Melrose Park, IL 60160
(708)450-3000

♥ Alexander de Markoff, Ltd.
(see Revlon)

♥ Alexander, Inc. ☆ ■
25030 Avenue Tibbitts,
Suite P
Valencia, CA 91355-3437
Cosmetics, Personal care

♥ Alexandra Avery Purely
Natural Body Care ☆ ■ ✉
4717 SE Belmont
Portland, OR 97251
(800)669-1863,
(503)236-5926
Personal care, Cosmetics

♥ Alfin, Inc.
(see Adrien Arpel)

♥ Aliage
(see Estee Lauder)

♥ Alivio Products Inc. ☆ ■
17150 E. Ohio Drive
Aurora, CO. 80017
(303)745-9900
Cosmetics

▼ All Detergents
(see Lever Brothers)

♥ All Heal Salve
(see Wise Ways Herbals)

♥ All-One-God Faith, Inc. ☆ ■
P.O. Box 28
Escondido, CA 92033
(619)745-7069,
(619)743-2211
Personal care

♥ All-Nutrient
(see Chuckles)

♥ Allens Naturally ☆ ■ ● ✉
P.O. Box 339, Dept. T
Farmington, MI
48332-0339
(800)352-8971,
(313)453-5410
Household

♥ Allercreme Products
(see Carme)

☆ Products contain no   ■ Manufacturer ● Distributor ✉ Mail Order
animal or animal
derived ingredients

♥ Allerpet, Inc. ☆ ■
P.O. Box 1076
Lenox Hill Station
New York, NY 10021
(212)861-1134

♥ Almay, Inc.
(see Revlon)

♡ Almond Glow
(see Home Health Products)

♥ Aloe 99
(see Alivio Products)

♥ Aloe Baby
(see Alivio Products)

♥ Aloe Burst
(see Aloe Up, Inc.)

♥ Aloe Chamomile
(see Carme)

♥ Aloe Creme
Laboratories ☆ ■
738 Union Avenue
Middlesex, NJ 08846
(800)327-4969,
(908)563-0077
Cosmetics, Personal care

♥ Aloe E
(see Alivio Products)

♥ Aloe Flex Products,
Inc. ☆ ■
P.O. Drawer 1347
Dickinson, TX 77539
(800)231-0839,
(713)337-2240
*Cosmetics, Personal
care, Companion animal*

♥ Aloe Gold
(see Green Mountain)

♥ Aloe Ice
(see Aloe Up, Inc.)

♥ Aloe Jojoba
(see Carme)

♥ Aloe Kote
(see Aloe Up, Inc.)

♡ Aloe Soap
(see Botanicus)

♥ Aloe Spa
(see Aloette Cosmetics, Inc.)

♥ Aloe Up, Inc. ■
P.O. Box 2913
Harlingen, TX 78551
(800)537-ALOE,
(210)428-0081
Cosmetics, Personal care

♥ CRUELTY FREE
Does not test products
or ingredients on animals

♥ Ingredients MAY
be tested on
animals

▼ Tests products or
ingredients
on animals

♥ Aloe Vera Gel Plus Herbs
(see Wachters)

♥ Aloegen
(see Carme)

♥ Aloegen Natural Cosmetics
(see Levlad, Inc.)

♥ Aloeprime Inc. ■ ●
10575 Newkirk, Suite 780
Dallas, TX 75220
*Cosmetics, Household,
Personal care, Companion
animal*

♥ Aloette Cosmetics, Inc. ■ ●
1301 Wright's Lane East
West Chester, PA 19380
(800)321-ALOE, (610)692-0600
Cosmetics, Personal care

♥ Alpen Limited ☆ ■
11 Pratt Street
Essex, CT 06426
(203)767-2862
Companion animal

♥ Alpha 9/JDS Manufac-
turing Co., Inc. ☆ ■
7718 Burnet Ave.
Van Nuys, CA 91405
(818)994-9599
Personal care

▼ Alpha Keri
(see Bristol-Myers Squibb)

♥ Alpha-Frutein
(see Palm Beach Beauty
Products)

♡ Alvin Last, Inc. ■
19 Babcock Place
Yonkers, NY 10701-2714
(800)527-8123,
(914)376-1000
Cosmetics, Personal care

▼ Always
(see Procter & Gamble)

♥ Always Save
(see Associated Whole-
sale Grocers)

♥ Amazing Clean
(see Minto Industries Ltd.)

♥ Amazing Grains
(see Body Love Natural
Cosmetics)

♥ Amazon Premium
Products ☆ ■
P.O. Box 530156
Miami Shores, FL 33153
(800)832-5645
Household

☆ Products contain no   ■ Manufacturer   ● Distributor   ⊠ Mail Order
animal or animal
derived ingredients

♥ Amber Essence
(see Eden Botanicals)

♥ Amber Rosewood
(see Eden Botanicals)

♥ Amberwood ☆ ✉
Route 2, Box 300, Milford
Road, Baker County
Leary, GA 31762
(912)792-6246
*Cosmetics, Personal care,
Companion animal, Household*

▼ Ambi Skin Care Products
(see Kiwi Brands Inc.)

♥ Amera Sales, Inc. ■ ●
6403 West Grandridge Blvd.
Kennewick, WA 99336
(509)735-1531
Cosmetics, Personal care

♥ America's Finest
Products Corp. ☆ ■
1639-9th Street
Santa Monica, CA 90404
(310)450-6555
Household

♥ American Cosmetics
Industries ☆ ■
P.O. Box 1309
Torrance, CA 90503
(310)214-1811
Cosmetics

♥ American Eco-Systems ☆ ■
125 9th Street
Wellman, IA 52356
*Household, Companion
animal*

♥ American International
Ind. ☆ ■
2220 Gaspar Avenue
Los Angeles, CA 90040
(213)728-2999
Cosmetics, Personal care

♥ American Merfluan, Inc.
(see Eco-Dent Interna-
tional, Inc.)

♥ American ORSA, Inc. ☆ ■
75 North St., P.O. Box 219
Redmond, UT 84652
(800)367-7258
Personal care

❤ Ampro Industries ☆ ■
850 N. Old US 23
Brighton, MI 48116
(810)632-5640
Companion animal

❤ Amrita
(see Auromere Ayurvedic
Imports)

❤ Amway Corporation ■
7575 Fulton Street, East
Ada, MI 49355-0001
(616)676-6000
Cosmetics, *Personal care*,
Household

♡ Anais Anais
(see L'Oreal of Paris)

❤ Ananda Country
Products ☆ ■ ● ✉
14618 Tyler Foote Road
Nevada City, CA 95959
(800)537-8766,
(916)292-3505
Personal care

❤ Ananda Incense
(see Ananda Country
Products)

❤ Ancient Formulas Inc. ■ ●
P.O. Box 1313
Wichita, KS 67201
(800)543-3026,
(316)634-2000
*Personal care, Companion
animal*

❤ Andalina, Ltd. ■ ✉
Tory Hill
Warner, NH 03278-0057
(603)456-3289
Personal care

❤ Andrew Jergens
Company ■
P.O. Box 145444
Cincinnati, OH 45250
(513)421-1400
Personal care

❤ Anew
(see Avon Products)

▼ Angel Face
(see Chesebrough-Pond's)

♡ Angel Soft Paper Products
(see Georgia-Pacific)

▼ Angus Chemical Company
1500 E. Lake Cook Road
Buffalo Grove, IL 60089-6556
(708)215-8626

☆ Products contain no   ■ Manufacturer   ● Distributor   ✉ Mail Order
   animal or animal
   derived ingredients

♥ Annemarie Borlind
Caring Color Collection
(see Borlind of Germany)

♥ Anti-Terge Cream & Lotion
(see Comfort Manufacturing)

▼ Aphrodisia Fragrances
(see Chesebrough-Pond's)

♥ Aphrodisia Products, Inc. ■
62 Kent Street
Brooklyn, NY 11222
(718)383-3677
Personal care

♥ Appearance
(see RC International)

▼ Apple Pectin
(see DowBrands)

♥ Apricot Boby Scrub
(see Wachters')

♡ Aqua de Selva
(see Mem Company)

▼ Aqua Fresh
(see SmithKline Beecham)

▼ Aqua Net
(see Chesebrough-Pond's)

▼ Aqua Velva
(see SmithKline Beecham)

♥ Aqualin Skin Moisturizer
(see Micro Balanced
Products)

♥ Aramis Inc.
(see Estee Lauder)

♥ Arbonne
International ☆ ■ ●
15 Argonaut
Aliso Viejo, CA 92656
(800)ARBONNE
(714)770-2610
Cosmetics, Personal care

♥ Arizona Gold, Oil of Jojoba
(see Jojoba Resources)

♥ Arizona Natural
Resources, Inc. ■
2525 East Beardsley Road
Phoenix, AZ 85024
(602)569-6900
*Cosmetics, Personal care,
Household*

▼ Arm & Hammer
(see Church & Dwight)

♡ Armani
(see L'Oreal of Paris)

▼ Armstrong Cleaner
(see S.C. Johnson & Son)

▼ Armstrong Floor Cleaners
(see Reckitt & Colman)

♥ Aroma Essence
(see Desert Essence)

♥ Aroma Life Company ☆ ■
P.O. Box 7371
Van Nuys, CA 91409
(805)944-4909
Cosmetics, Personal care

♥ Aroma Vera, Inc. ☆ ■
5901 Rodeo Road
Los Angeles, CA 90016
(800)669-9514, (310)280-0407
Personal care

♥ Aromaland, Inc. ☆ ■ ●
Route 20, Box 29 A.L.
Santa Fe, NM 87501
(800)933-5267,
(505)438-0402
Personal care

♥ Aromalotion
(see Body Love Natural
Cosmetics)

♥ Aromatherapy for Kids
(see La Natura)

♥ Associated Wholesale
Grocers, Inc. ●
P.O. Box 2932
Kansas City, KS 66110-2932
(913)321-1313
Personal care

▼ Atra
(see Gillette Company)

♥ Atta Lavi ☆ ■ ✉
443 S. Oakhurst Dr., #305
Beverly Hills, CA 90212
Personal care

♥ Attar Bazarr ☆ ■
(see Chishti Company)

▼ Attends
(see Procter & Gamble)

♥ Attogram Corporation
(see Acu-Trol, Inc.)

♥ Aubrey Organics ■
4419 North Manhattan Ave.
Tampa, FL 33614
(800)282-7394, (813)877-4186
*Cosmetics, Personal care,
Household, Companion
animal*

☆ Products contain no   ■ Manufacturer   ● Distributor   ✉ Mail Order
animal or animal
derived ingredients

♡ Aura Cacia, Inc. ☆ ■
P.O. Box 399
Weaverville, CA 96093
(800)437-3301, (916)623-3301
Personal care

♥ Auroma International, Inc.
(see Lotus Light Ent.)

♥ Auromere Ayurvedic
Imports ☆ ✉
1291 Weber Street
Pomona, CA 91768
(800)735-4691,(909)629-0108
Personal care

♥ Aurora Henna Company ■
1507 East Franklin Avenue
Minneapolis, MN 55404
(612)870-4456
Personal care

♡ Aussie
(see Redmond Products)

♥ Austin Diversified Products
(see Advantage Wonder
Cleaner)

♥ Austin's ☆ ■
P.O. Box 827
Mars, PA 16046-0827
(800)245-1942, (412)625-1535
Household

♥ Autumn Harp, Inc. ■
P.O. Box 267
Bristol, VT 05443
(802)453-4807
Personal care

♥ Avalon Natural
Cosmetics, Inc. ☆ ■ ●
P.O. Box 750428
Petaluma, CA 94975-0428
(707)769-5120
Cosmetics, Personal care

♥ Avanza Corp. ☆ ■
11818 San Marino Street
Rancho Cucamonga, CA
91730
(909)944-3447
Cosmetics, Personal care

♥ Aveda Corporation ■
4000 Pheasant Ridge Drive
Blaine, MN 55449
(800)328-0849, (612)783-4000
*Cosmetics, Personal care,
Household*

▼ Aveeno
(see S.C. Johnson & Son)

▼ Aviance
(see Chesebrough-Pond's)

♥ CRUELTY FREE
Does not test products
or ingredients on animals

♥ Ingredients MAY
be tested on
animals

▼ Tests products or
ingredients
on animals

♡ Avigal Henna ■ ● ⊠
45-49 Davis Street
Long Island City, NY 11101
(800)722-1011
Cosmetics, Personal care

♥ Avon Products, Inc. ■ ●
9 W. 57th Street
New York, NY 10019-2683
(800)445-AVON,
(212)546-6015
Cosmetics, Personal care

♥ Aware Diaper, Inc. ☆ ■
P.O. Box 2591
Greeley, CO 80632
(303)352-6822
Household

▼ Axion Detergent Booster
(see Colgate Palmolive)

♥ Ayurherbal Corp.
(see Lotus Light Ent)

♥ Ayurveda Holistic
Center ☆ ■ ● ⊠
82A Bayville Avenue
Bayville, NY 11709
(516)628-8200
Personal care

▼ Aziza
(see Chesebrough-Pond's)

♥ Aztec Secret ☆ ■ ●
P.O. Box 19735
Las Vegas, NV 89132
(702)386-5680
Personal care

♥ Azur Fragrances USA,
Inc. ☆ ■
205 E. 60th Street.
New York, NY 10022-1434
Personal care

♥ B.J. Griffing Luxury Dog
Shampoo Bar
(see New Direction)

▼ Babe
(see Chesebrough-Pond's)

▼ Baby Bath
(see Mennen)

▼ Baby Magic
(see Mennen)

♥ Baby Massage
(see Green Mountain)

♥ Baby Mild + Kind
Children Products
(see Borlind of Germany)

♥ Back to Basics
(see Smith & Vandiver)

☆ Products contain no   ■ Manufacturer   ● Distributor   ⊠ Mail Order
animal or animal
derived ingredients

💜 Bailey Group Samuel
Par Paris ☆ ● ✉
7760 Romaine Street
West Hollywood, CA 90046
(213)654-3301
Cosmetics, Personal care

💜 Baja Beach ☆ ■
16016 S. Figueroa Street
Gardena, CA 90248-2435
(800)875-4TAN, (310)374-4928
Cosmetics

▼ Balm Barr Cocoa Butter
(see Mennen)

▼ Balsam Color
(see Clairol)

▼ Ban
(see Bristol-Myers Squibb)

♡ Banana Boat
(see Sun Pharmaceuticals)

💜 Band Ade
(see Minto Industries Ltd.)

▼ Band-Aid
(see Johnson & Johnson)

▼ Banner
(see Procter & Gamble)

♡ Bar Keepers Friend
(see SerVaas Labs.)

💜 Barbizon International,
Inc. ✉
2240 SW 15th Avenue
Suite 300
Boynton Beach, FL
33426-6363
(407)362-8883
Cosmetics, Personal care

▼ Bare Elegance
(see Gillette)

💜 Bare Escentuals ☆ ■
#600 Townsend Street
Suite 329 East
San Francisco, CA 94103
(800)227-3990,
(415)487-3400
Cosmetics, Personal care

▼ Baron
(see Unilever United States)

💜 Barristo, Ltd. ☆ ■ ●
600 N. McClurg Court
Chicago, IL 60611
(312)454-1214
Personal care

❤ Basic Elements Hair
Care System, Inc. ☆ ●
505 S. Beverly Dr.
Suite 1292
Beverly Hills, CA 90212
(800)947-5522, (310)586-1955
Personal care

▼ Basic White
(see Clairol)

❤ Basically Natural ☆ ✉
109 East G Street
Brunswick, MD 21716
(800)352-7099, (301)834-7923
*Cosmetics, Personal care,
Household, Companion
animal*

❤ Bath & Body Works
(see Gryphon Development)

❤ Bath & Shower Gelee
(see Wachters')

❤ Bath Buddies Natural
Bath Products For Children
(see Wellington Labs)

❤ Bath Island, Inc. ✉
469 Amsterdam Ave.
New York, NY 10024
(212)787-9415
Cosmetics, Personal care

▼ Bath Talc
(see Mennen)

❤ Bath Therapy
(see Para Labs)

❤ Bathmoods
(see Levlad, Inc.)

❤ Baudelaire, Inc. ✉
Forest Road
Marlow, NH 03456
(800)327-2324, (603)352-9234
Personal care

▼ Bausch & Lomb, Inc.
P.O. Box 450
Rochester, NY 14692
(800)344-8815

❤ Bavarian Alpenol &
Sunspirit ☆ ■ ●
1343 N. Nevada Ave
Colorado Springs, CO 80903
(719)633-8931
Personal care, Cosmetics

❤ Beaumont Products,
Inc. ☆ ■
2197 Canton Road, Ste. 201
Marieta, GA 30066
(800)451-7096,
(404)514-9600
Household, Personal care

☆ Products contain no   ■ Manufacturer   ● Distributor   ✉ Mail Order
animal or animal
derived ingredients

34

♥ BeautiControl
Cosmetics, Inc. ■
2121 Midway Road
Carrollton, TX 75006
(800)326-2365, (214)458-0601
Cosmetics, Personal care

▼ Beautiflor
(see S.C. Johnson & Son)

♥ Beautiful
(see Estee Lauder)

♥ Beautiful Belly Balm
(see Wise Ways Herbals)

▼ Beautiful Collection
(see Clairol)

♥ Beauty Drops
(see Colonial Dames)

♥ Beauty For All Seasons ● ✉
P.O. Box 51810
Idaho Falls, ID 83405-1810
(800)942-4336
Cosmetics, Personal care

♥ Beauty Naturally ☆ ■ ✉ ●
P.O. Box 4905
Burlingame, CA 94010
(800)432-4323, (415)697-7547
Personal care

♥ Beauty Time, Inc. ■
299 McMurray Rd.
Upper St. Claire, PA 15241
Cosmetics, Personal care

♥ Beauty Without Cruelty
Cosmetics
(see Avalon Natural
Cosmetics)

♥ BeautyMasters ☆ ■ ●
HCR 68, Box 745
Vian, OK 74962
(918)489-5164
Cosmetics, Personal care

♥ Becoming Cologne
(see BeautiControl
Cosmetics)

♥ Becoming Color
(see BeautiControl Cosmetics)

♥ Beehive Botanicals Inc. ■
Box 8257
Hayward, WI 54843
(800)283-4274, (715)634-4274
Personal care

▼ Befresh
(see S.C. Johnson & Son)

▼ Begin Again Conditioner
(see DowBrands, Inc.)

♥ CRUELTY FREE
Does not test products
or ingredients on animals

♥ Ingredients MAY
be tested on
animals

▼ Tests products o
ingredients
on animals

▼ Behold
(see Kiwi Brands)

💜 Beiersdorf, Inc. ■
360 Martin Luther King Dr.
Norwalk, CT 06856-5529
(800)227-4703
Cosmetics, Personal care

💜 Bella's Secret Garden ■
P.O. Box 2189
Karrel Islands, CA 93034
(800)962-6867, (805)483-5750
Personal care, Household

💜 Belle Star Inc. ■ ✉
23151 Alcalde, #C4X
Laguna Hills, CA 92653
(800)442-STAR,
(714)768-7006

💜 Ben Nye Makeup ☆ ■
5935 Bowcroft Street
Los Angeles, CA 90016
(310)839-1984
Cosmetics

💜 Benetton Cosmetics
Corporation ■
540 Madison Ave., 29th Flr.
New York, NY 10022
(212)832-6616
Cosmetics, Personal care

▼ Benjamin Ansehl Company
1555 Page Industrial Blvd.
St. Louis, MO 63132
(800)937-2284,
(314)429-4300

▼ Benzodent
(see Procter & Gamble)

💜 Best Choice
(see Associated Whole-
sale Grocers)

💜 Beverly Hills Spa
(see Conair)

▼ Beyond Time
(see Marche Image)

💜 Bi-O-Kleen Industries
Inc. ☆ ■
P.O. Box 82066
Portland, OR 97282-0066
(503)774-1295
Household

💜 Bianca of Beverly Hills
(see Earth Safe)

▼ Big Body
(see Gillette)

☆ Products contain no   ■ Manufacturer   ● Distributor   ✉ Mail Order
    animal or animal
    derived ingredients

💜 Bilange Inc. ☆ ●
227 E. 56th Street, Flr. 201
New Yrok, NY 10022-3754
(800)477-6643
Personal care

💜 Bill Blass
(see Revlon)

💜 Bio Nature Internation
(see Martin Von Myering)

♡ Bio Sentry Labs ☆ ■
4600 Park Avenue
Des Moines, IA 50321
(515)243-3000
*Cosmetics, Household,
Personal care, Companion
animal*

💜 Bio-Botanica, Inc. ☆ ■
75 Commerce Drive
Hauppauge, NY 11788
Cosmetics, Personal care

💜 Bio-Pac ☆ ■
P.O. Box 580
Union, ME 04862
(207)785-2855
Household

💜 Bio-Tec Cosmetics
Ltd. ☆ ■
92 Sherwood Avenue
Toronto, Ontario,
Canada M4P 2A7
(800)667-2524
Personal care

💜 Bio-Tec Professional
Haircare Products
(see Bio-Tec Cosmetics)

💜 BioAdvance
(see Avon Products)

💜 BIOCLEAN
(see Natural Chemistry)

💜 Bioessence International
Labs
(see Essential Products of
America)

💜 BioFilm, Inc. ☆ ■
3121 Scott Street
Vista, CA 92083
(800)848-5900, (619)727-9030
Personal care

💜 Biogena
(see Conair)

💜 Biogime
(see Entourage/Biogime)

♥ Biokosma
(see Caswell-Massey)

♡ Biopractic Group II, Inc. ☆ ■
P.O. Box 5300
Phillipsburg, NJ 00865
(908)859-4060
Cosmetics

♥ Biotene H-24
(see Carme)

♡ Biotherm
(see L'Oreal of Paris)

▼ Biz
(see Procter & Gamble)

▼ Black Flag
(see Reckitt & Colman)

♥ Black Pearl Gardens ■ ● ⊠
425 S. Main Street
Franklin, OH 45005
(800)891-0142, (513)746-0004
Personal care

♥ Blackmores Ltd. ■ ●
16 Parkside Drive
North Brunswick, NJ
08902-1218
(908)422-4888
Personal care

♥ Blessed Herbs ☆ ⊠
109 Barre Plains Rd
Oakham, MA 01068
(508)882-3839
Personal care

♥ Block & Tackle Sunblock
(see Finley Companies)

▼ Block Out-SPF
(see Chesebrough-Pond's)

▼ Blue Blades
(see Gillette)

♡ Blue Coral, Inc. ☆ ■
5300 Harvard Ave.
Cleveland, OH 44105
(216)351-3000
Household

♥ Blue Cross Beauty
Products, Inc. ☆ ■
12251 Montague Street
Pacoima, CA 91331
(818)896-8681
Cosmetics, Personal care

♥ Blue Pearl
(see Siddha Int'l)

💜 Blue Ribbons Pet Care ☆ ✉
2475 Bellmore Avenue
Bellmore, NY 11710
(800)552-BLUE
Companion animal

💜 Bo-Chem
(see Neways, Inc.)

💜 Bob Kelly Cosmetics ☆ ◾
151 West 46th Street
New York, NY 10036
(212)819-0030
Cosmetics

💜 Bocabelli Inc. ☆ ◾ ✉ ●
5539 Erie Ave., N.W.
Canal Fulton, OH 44614
(216)477-9048
Personal care

💜 Body & Massage Lotion
(see Free Spirit Ent.)

💜 Body & Soul of Chicago ● ✉
300 West Grand
Chicago, IL 60610
(800)272-7085
Cosmetics, Personal care

💜 Body Encounters ✉
604 Manor Road
Cinnaminson, NJ 08077
(609)829-4660
Personal care

💜 Body Guard Pet
Food Supplement
(see Pro-Tec Pet Health)

💜 Body Lind Body Care &
Cellulite Cream
(see Borlind of Germany)

💜 Body Love Natural
Cosmetics, Inc. ☆ ◾
P.O. Box 7542
303 Petrero Street, #19
Santa Cruz, CA 95061-7542
(408)425-8218
Personal care

💜 Body Shop, Inc. ✉ ●
One World Way
Wake Forest, NC 27987
(919)554-4900
Cosmetics, Personal care

💜 Body Suite ☆ ◾ ✉
316 Manhattan Beach Blvd.
Manhattan Beach, CA 90266
(800)541-2535, (310)379-4840
*Cosmetics, Personal care,
Companion animal*

♡ Body Therapies
(see Orjene Natural
Cosmetics)

♥ Body Time ✉
1341 Seventh Street
Berkeley, CA 94710
(510)524-0216
Personal care

♥ Body Tools ■ ✉
16 Pamaron Way, Suite C
Novato, CA 4949-6217
Personal care

♥ Body/Body Bath &
Body Products
(see Avalon Natural Cosmetics)

♥ Bodyography ■ ✉ ●
1641 16th Street
Santa Monica, CA 90404
(800)642-BODY
(310)399-2886
Cosmetics, Personal care

▼ Bold
(see Procter & Gamble)

▼ Bold Hold
(see Alberto-Culver)

♥ Bon Sante
(see Carme)

♥ Bonne Bell, Inc. ☆ ■ ●
18519 Detroit Avenue
Lakewood, OH 44107
(800)321-1006,
(216)221-0800
Cosmetics, Personal care

♥ Borax
(see Dial Corporation)

♥ Borghese
(see Revlon)

♥ Borlind of Germany, Inc. ■
P.O. Box 130
New London, NH 03257
(800)447-7024, (603)526-2076
Cosmetics, Personal care

♥ Botan Corporation ☆ ■ ● ✉
7760 Romaine Street
W. Hollywood, CA 90046
(800)448-0800, (213)654-3301
Cosmetics, Personal care

♥ Botanical Products, Inc. ■
34725 Bogart Drive
Springville, CA 93265
(209)539-3432
Personal care

♥ Botanicals
(see Smith & Vandiver)

☆ Products contain no
animal or animal
derived ingredients   ■ Manufacturer   ● Distributor   ✉ Mail Order

💜 Botanics of California ☆ ■
3001 South State, #29
Ukiah, CA 95482
(707)462-6141
Cosmetics, Personal care

💜 Botanics Skin Care
(see Botanics of California)

♡ Botanicus Inc. ☆ ■ ●
7610 Rickenbacker Dr., Ste. T
Gaithersburg, MD 20879
(800)282-8887,
(301)977-8887
Personal care

▼ Bounce
(see Procter & Gamble)

▼ Bounty
(see Procter & Gamble)

💜 Bradford Soap Works, Inc. ■
P.O. Box 1007
West Warwick, RI 02893
(401)821-2141
Personal care

▼ Brasso Metal Cleaners
(see Reckitt & Colman)

▼ Braun
(see Gillette)

💜 Breck Hair Care
(see Dial Corporation)

▼ Breeze
(see Lever Brothers)

💜 Breezy Balms ☆ ■
2553A Mission Street
Santa Cruz, CA 95060
(408)688-8706
Personal care

💜 Brillo Pads
(see Dial Corporation)

▼ Bristol-Myers Squibb
Company
345 Park Avenue
New York, NY 10154
(212)546-4000

▼ Brita
(see Clorox)

▼ Brite
(see S.C. Johnson & Son)

💜 Bronson Pharmaceuticals
1945 Craig Road
St. Louis, MO 63146
(800)235-3200
Personal care

💜 CRUELTY FREE
Does not test products
or ingredients on animals

💜 Ingredients MAY
be tested on
animals

▼ Tests products or
ingredients
on animals

❤ Brooks
(see Traditional Products)

❤ Brookside Soap
Company ☆ ■ ● ✉
P.O. Box 55638
Seattle, WA 98155
(206)742-2265
*Personal care, Companion
animal*

▼ Brow Tamer
(see Max Factor)

❤ Brush Craft ■
P.O. Box 1647
Clifton, NJ 07015
(201)779-8800
Personal care

❤ Brush On Nail Gel System
(see Gena Labs)

▼ Brush Plus Shaving System
(see Gillette)

▼ Brut Products
(see Chesebrough-Pond's)

▼ Brylcreem
(see SmithKline Beecham)

❤ Bufette
(see Delby System)

❤ Bug-Off, Inc. ☆ ■ ✉ ●
Rt2, Box 248C
Lexington,VA 24450
(703)463-1760
*Personal care, Household,
Companion animal*

▼ Bully Toilet Bowl Cleaner
(see Reckitt & Colman)

▼ Burberry's for Men
(see Unilever U.S.)

❤ Bygone Bugs, Herbal
Outdoor Formula
(see Lakon Herbals)

❤ C & S Labs
(see Sombra Cosmetics Inc.)

❤ C.E. Jamieson & Co., Ltd. ■
4025 Rhodes Drive
Windsor, Ontario
Canada N8W 5B5
(519)974-8482
*Cosmetics, Personal care,
Companion animal*

❤ Cabot Labs ☆ ■
165 Oval Drive
Central Islip, NY 11722
(800)645-5048
Cosmetics, Personal care

☆ Products contain no   ■ Manufacturer   ● Distributor   ✉ Mail Order
animal or animal
derived ingredients

♡ Cacharel
(see L'Oreal of Paris)

▼ Cachet
(see Chesebrough-Pond's)

▼ Calgon Corporation
P.O. Box 1346
Pittsburgh, PA 15230
(412)777-8000

▼ California
(see Max Factor)

♥ California Beauty Bar
(see Action Labs)

♥ California Colors
(see Blue Cross Beauty
Products)

♥ California Gold
(see J & J Jojoba)

♥ California Naturals
(see California Olive Oil)

♥ California Olive Oil
Corp. ☆ ■ ● ✉
134 Canal Street
Salem, MA 01970
(508)745-7840
Personal care, Household,
Companion animal

♥ California Shine
(see Conair)

♥ California Skin Therapy ☆ ■
270 North Canon Dr., #1297
Beverly Hills, CA 90210
(800)SUN-CARE,
(310)824-2508
Personal care

♥ California Tan ☆ ■
(see California Skin Therapy)

▼ Camay
(see Procter & Gamble)

♥ Cameo Copper
(see Dial Corporation)

♥ CamoCare
(see Abkit)

♥ Canada's All Natural
Soap, Inc. ■
Unit 9, 1211 Gorham Street
Newmarket, Ontario
Canada L3Y 7V1
(416)853-2677
Cosmetics, Personal care

● Canterbury Chemists ✉
1020 Lake Street
Oak Park, IL 60301
(708)848-7999
Cosmetics, Personal care

● Cara Bella Natural Cosmetics
(see Royal Labs)

● Carbona Products
Company ☆ ■ ●
17-20 Whitestone Expy, #201
Whitestone, NY 11357-3000
Household

● Card Corporation ☆ ■
2513 Elmira Street
Aurora, CO 80010
(303)739-9614
Household

▼ Care Bear Tissues
(see Kimberly-Clark)

▼ Carefree Panty Shields
(see Johnson & Johnson)

▼ Caress
(see Lever Brothers)

● Caribbean Pacific of the
Rockies ☆ ■ ●
P.O. Box 380
Crawford, CO 81415
(303)921-4050
Personal care

● Carina Supply Inc. ☆ ■
464 Granville Street
Vancouver, BC, Canada
V6C 1V4
(800)663-0479, (604)687-3617
*Personal care, Companion
animal*

● Caring Catalog ✉
7678 Sagewood Drive
Huntington Beach, CA 92648
(714)842-0454
*Cosmetics, Personal care,
Household, Companion
animal*

♡ Carlson
(see J.R. Carlson Labs)

● Carma Labs, Inc. ■
5801 W. Airways Avenue
Franklin, WI 53132
(414)421-7707
Personal care

♥ Carme, Inc. ■
84 Galli Drive
Novato, CA 94949
(800)227-2628, (415)382-4000
Cosmetics, Personal care

♥ Carmex
(see Carma Labs.)

▼ Carol Richards
(see Unilever U.S.)

♥ Caroline Cares ☆ ✉
4703 Delmar
Rockford, IL 61108
(815)397-3341
Personal care

▼ Carpet Fresh
(see Reckitt & Colman)

♥ Carrington Parfums
(see Revlon)

▼ Cascade
(see Procter & Gamble)

▼ Cashmere Bouquet
(see Colgate-Palmolive)

♥ Cassini Parfums, Ltd. ■ ●
3 West 57th Street, 8th Flr
New York, NY 10019
(800)441-6534
Cosmetics

♥ Caswell-Massey ● ✉
121 Fieldcrest Avenue
Edison, NJ 08818
(908)225-2181
Personal care, Cosmetics

♥ Cat's Pride Cat Litter
(see Oil-Dri)

♥ Catalog Ventures, Inc. ✉
27 Industrial Avenue
Chelmsford, MA 01824
(508)256-4100
Personal care

▼ Cavalier America Inc.
(see Kiwi Brands)

♥ Celebrity
(see Studio Magic)

♥ Cellagen
(see Earth Science)

♡ Century Systems ■
120 Selig Drive, SW
Atlanta, GA 30336
(800)843-96626
Cosmetics

♥ Cernitin America Inc. ☆ ✉ ●
P.O. Box 261193
Columbus, OH 43226
(800)237-6426, (614)842-4288
Personal care

♥ CRUELTY FREE
Does not test products
or ingredients on animals

♥ Ingredients MAY
be tested on
animals

▼ Tests products or
ingredients
on animals

♥ Chae
(see Alivio Products)

♡ Chamovera
(see Home Health Products)

♥ Chandrika
(see Auromere Ayurvedic
Imports)

♥ Chanel, Inc. ■
Nine West 57th Street
New York, NY 10019
(212)688-5055
Cosmetics, Personal care

♥ Change of Face
Cometics ☆ ■ ✉
8313 Calumet Avenue
Munster, IN 46321
(219)947-4040
Cosmetics, Personal care

♥ Changes Natural ☆ ■ ✉
7400 San Pedro Avenue,
Suite 1250
San Antonio, TX 78216-5357
(512)341-1709
Cosmetics, Personal care

♡ Chantal Skin Care
(see National Home Care
Products)

♥ Chanty Inc. ●
351 36th Street
Brooklyn, NY 11232
Cosmetics, Personal care

♥ Charade Cologne
(see BeautiControl
Cosmetics)

♥ Charles of the Ritz
(see Revlon)

♥ Charlie
(see Revlon)

▼ Charmin
(see Procter & Gamble)

♥ Chas Russell
(see LaDove)

▼ Chase
(see Bristol-Myers Squibb)

♥ Chatoyant Pearl
Cosmetics ■
P.O. Box 526
Port Townsend, WA
98368
(206)385-4825
Cosmetics, Personal care

♥ Chaz
(see Revlon)

☆ Products contain no   ■ Manufacturer   ● Distributor   ✉ Mail Order
   animal or animal
   derived ingredients

▼ Cheer
(see Procter & Gamble)

♡ Chemifax
(see Blue Coral)

♥ Chempoint Products, Inc. ☆ ■
P.O. Box 2597
Danbury, CT 06813-2597
(800)343-6588,
(203)778-0881
Household, Personal care

♥ Chenti Products, Inc. ■
21093 Forbes Avenue
Hayward, CA 94545
(510)785-2177
Cosmetics, Personal care

▼ Chesebrough-Pond's USA
33 Benedict Place
Greenwich, CT 06830
(203)661-2000

♥ Chica Bella, Inc. ■
1055 17th Ave.
Santa Cruz, CA 95062
(408)462-1280
Personal care

▼ Chimere
(see Prince Matchabelli)

♥ CHIP Distribution
Company ☆ ■
8321 Croydon Avenue
Los Angeles, CA 90045
(310)545-8933
Household

♥ Chishti Company ☆ ■
79 New Virginia Road
Oxford, NY 13830
(800)344-7172, (607)843-5166
Personal care

▼ Chloe
(see Unilever U.S.)

♥ Chloro Sea Creme
(see Wachters')

▼ Chore Boy
(see Reckitt & Colman)

♥ Christian Dior Perfumes,
Inc. ■ ●
9 West 57th Street
New York, NY 10019
(212)759-1840
Cosmetics, Personal care

♥ Christine Valmy ☆ ■
285 Change Bridge Rd.
Pine Brook, NJ 07058
(800)526-5057, (201)575-1050
Cosmetics

♥ CRUELTY FREE
Does not test products
or ingredients on animals

♥ Ingredients MAY
be tested on
animals

▼ Tests products or
ingredients
on animals

♡ Christopher Enterprises,
Inc. ■ ● ⊠
P.O. Box 777
Springville, UT 84663
(801)489-8787
Personal care

♥ Chuckles, Inc. ☆ ■ ●
P.O. Box 5126
Manchester, NH 03108-5126
(800)221-3496, (603)669-4228
Personal care

▼ Church & Dwight Co., Inc.
469 North Harrison Street
Princeton, NJ 08543-5297
(800)524-1328, in NJ,
(800)624-2889

♥ CiCi Cosmetics ■ ●
9500 Jefferson Blvd.
Culver City, CA 90232
(800)869-1224, (310)815-1476
Cosmetics

♥ Cielo
(see Professional &
Technical Services)

♥ Cinema Secrets, Inc. ■ ● ⊠
4400 Riverside Drive
Burbank, CA 91505
(818)846-0579
Cosmetics, Personal care

♥ Cinnabar
(see Estee Lauder)

♥ Citra Brands, Inc. ☆ ■
12555 Enterprise Blvd.
Largo, FL 34643
(800)36CITRA, (813)535-3459
Household, Personal care

♥ Citra-Glow
(see Mia Rose Products)

♥ Citra-Solv
(see Chempoint Products)

♥ Citre Shine
(see Advanced Research Labs)

♥ Citrus II
(see Bascially Natural)

♥ Citrus Magic
(see Beaumont Products)

▼ Clairol, Inc.
345 Park Avenue
New York, NY 10154
(800)223-5800

♥ Clarins of Paris ☆ ■
135 East 57th Street, 15th Flr
New York, NY 10022
(212)980-1800
Cosmetics, Personal care

☆ Products contain no   ■ Manufacturer   ● Distributor   ⊠ Mail Order
animal or animal
derived ingredients

▼ Clarion
(see Procter & Gamble)

💜 Classic Cosmetics, Inc. ☆ ■
9601 Irondale Avenue
Chatsworth, CA 91311-5009
(818)773-9042
Cosmetics, Personal care

♡ Classic Form
(see Mem Company)

▼ Clean & Clear
(see Johnson & Johnson)

♡ Clean & Natural
(see Mem Company)

▼ Clear & Lovely
(see Colgate-Palmolive)

▼ Clear by Design
(see SmithKline Beecham)

▼ Clear Gel
(see Gillette)

💜 Clear Springs
(see Faith Products)

▼ Clearasil
(see Procter & Gamble)

💜 Clearly Natural
Products, Inc. ☆ ■
P.O. Box 750024
Petaluma, CA 94975-0024
(800)274-7627, (707)762-5815
Personal care

▼ Clearstick
(see Procter & Gamble)

💜 ClearVue Products, Inc. ☆ ■
P.O. Box 567
Lawrence, MA 01842
(508)794-3100
Household

💜 Cleopatra's Secret from
the Dead Sea ☆ ■ ●
130 W. 25th Street
New York, NY 10001
(800)344-9999, (212)645-4968
Personal care

💜 Clientele, Inc. ☆ ✉
5207 NW 163rd Street
Miami, FL 33014
(800)327-4660,
(305)624-6665
Cosmetics, Personal care

💜 Clinic
(see C.E. Jamieson & Co. Ltd.)

💜 CRUELTY FREE
Does not test products
or ingredients on animals

💜 Ingredients MAY
be tested on
animals

▼ Tests products or
ingredients
on animals

♡ Clinical-Formula
(see AHC Pharmacal)

♥ Clinique Laboratories Inc.
(see Estee Lauder)

▼ Clorox Company *Soft Scrub*
P.O. Box 24305 *Tilex*
Oakland, CA 94623-1305 *Clorox Toilet Cleaner*
(800)227-1860, *SOS + 409*
in CA (800)772-2469

▼ Close-up
(see Chesebrough-Pond's)

♥ Cloudworks ☆ ■
120 Tyler Road
Contoocook, NH 03229-3092
Personal care

▼ Coast
(see Procter & Gamble)

♥ Coastline Products ☆ ✉
P.O. Box 6397
Santa Ana, CA 92706
(800)554-4112
Personal care, Household

♥ Coat Guard Conditioning
Shampoo
(see Pro-Tec Pet Health)

▼ Cold Power
(see Colgate-Palmolive)

▼ Colgate-Palmolive
Company
300 Park Avenue
New York, NY 10022-7499
(800)338-8388

♥ Colin Ingram ☆ ■
61 Lloyden Drive
Atherton, CA 94027
(415)328-3184
Personal care

♥ Collagen Plus
(see Palm Beach Beauty
Products)

♥ Collection:Beginnings
(see Pacific Scents)

♥ Colonia, Inc. ☆ ■
4711 Bell Lane
Orange, CT 06477
Personal care

♥ Colonial Dames Co., Ltd. ■
P.O. Box 22022
Los Angeles, CA
90022-0022
(213)773-6441
Cosmetics, Personal care

💜 Color & Herbal Company ■
P.O. Box 5370
Newport Beach, CA 92662
(800)259-1701,
(714)721-1700
Companion animal

▼ Color Guard
(see DowBrands, Inc.)

💜 Color Me Beautiful ●
14000 Thunderbolt Place,
Suite E
Chantilly, VA 22021
(800)533-5503, (703)471-6400
Cosmetics, Personal care

💜 Color My Image ☆ ● ✉
5025B Backlick Rd.
Annadale, VA 22003
(703)354-9797
Cosmetics, Personal care

💜 Colora ☆ ■
217 Washington Avenue
Carlstadt, NJ 07072
(201)939-0969
Cosmetics

💜 Colorgraphics
(see Matrix Essentials)

▼ ColorHold
(see Clairol)

💜 Colourings/The Body
Shop ■ ● ✉
45 Horsehill Rd.
Cedar Knolls, NJ 07927
(201)984-9200
Cosmetics

💜 Columbia Cosmetics
Mfg., Inc. ☆ ■
1661 Timothy Drive
San Leandro, CA 94577
(800)824-3328,
(415)562-5900
Cosmetics, Personal care

💜 Comare Products ☆ ■ ●
5980 Miami Lakes Drive
Miami Lakes, FL 33014
*Personal care, Companion
animal*

▼ Combat
(see Clorox)

💜 Come To Your
Senses ☆ ■ ● ✉
321 Cedar Avenue, South
Minneapolis, MN 55454
Personal care

▼ Comet
(see Procter & Gamble)

❤ Comfort Manufacturing
Company ■
1056 West Van Buren Street
Chicago, IL 60607
(800)969-5573, (312)421-8145
Cosmetics, Personal care

❤ Common Scents ●
134 Main Street
Port Jefferson, NY 11777
(516)473-6370
Personal care

❤ Compassion Matters ☆ ● ✉
P.O. Box 3614
Jamestown, NY 14702-3614
(716)484-8502
Cosmetics, Personal care,
Household

❤ Compassionate Consumer ✉
P.O. Box 27
Jericho, NY 11753
(718)445-8566
Cosmetics, Personal care,
Household, Companion animal

▼ Complements
(see Clairol)

❤ Complexion Care Cleanser
(see Alexander Avery
Purely Natural Skin Care)

❤ Con-Lei
(see CHIP Distribution)

❤ Conair Corporation ■
1 Cummings Point Rd.
Stamford, CT 06904
(203)351-9173
Personal care

❤ Concept Now
Cosmetics ● ✉
P.O. Box 3208
Santa Fe Springs, CA 90670
(310)903-1450, (800)262-1215
Cosmetics, Personal care

▼ Condition
(see Clairol)

❤ Conditioner with Panthenol
(see Wachters')

▼ Consort
(see Alberto-Culver)

▼ Control Cat
(see Clorox)

❤ Cool Wash
(see Green Mountain)

▼ Cool Wave
(see Gillette)

❤ Coolove Cool Water Wash
(see America's Finest
Products)

❤ Cooper Development
(see Cabot Labs.)

♡ Copper Glo
(see SerVaas Labs.)

❤ Copper-Brite, Inc. ☆ ■
P.O. Box 50610
Santa Barbara, CA
93150-0610
Household, Companion animal

▼ Coppertone
(see Schering Plough)

❤ Cornucopia Valley Skin Care
(see Dep Corporation)

♡ Coronet Paper Products
(see Georgia-Pacific)

♡ Cosmair, Inc. ■
P.O. Box 98
Westfield, NJ 07091
(800)631-7358
Cosmetics, Personal care

❤ Cosmetic Source, Inc. ■
18 Grove Street, Apt. 7
Wellesley, MA 02181-7705
(508)658-7340
Cosmetics, Personal care

❤ Cosmetique ☆ ✉ ●
P.O. Box 9000
North Suburban, IL
60197-9939
(708)913-9099
Personal care

❤ Cosmo Cosmetics,
Inc. ☆ ■ ●
10455 Slusher Drive
Santa Fe Springs, CA
90670-3750
Cosmetics

❤ Cosmyl, Inc. ☆ ■
4401 Ponce De Leon Blvd.
Coral Gables, FL 33146
(305)446-5666
Cosmetics

❤ Cosmyl-Nail Tech
(see Cosmyl)

❤ Cot 'n' Wash ☆ ■
502 The Times Building
Ardmore, PA 19003
(215)896-4373
Household

▼ Coty, Inc
237 Park Avenue
New York, NY 10017-3142
(212)850-2300

♥ Country Comfort ■
28537 Nuevo Valley Drive
Nuevo, CA 92567
(800)462-6617,
(909)928-4038
Personal care

♥ Country Roads
(see Carme)

♥ Country Save
Corporation ☆ ■
3410 Smith Avenue
Everett, WA 98201
(206)258-1171
Household

♥ Coven Gardens,
Inc. ☆ ■ ✉ ●
P.O. Box 1064
Boulder, CO 80306
Cosmetics, Personal care

▼ Cover Girl
(see Procter & Gamble)

♡ Covermark Cosmetics
(Fomerly Lydia O'Leary
Cosmetics) ☆ ■ ● ✉
1 Anderson Avenue
Moonachie, NJ 07074
(201)460-7713
Cosmetics

♥ Crabtree & Evelyn, Ltd. ■ ✉
P.O. Box 158
Woodstock, CT 06281
(203)928-2761,
Outside CT (800)253-1519
Personal care

♥ Craig Martin
(see Comfort Manufacturing)

♥ Cream Polish
(see Venus Labs)

♡ Creamy Lime Bath Bubbles
(see House of Lowell)

♥ Creation Soaps,
Inc. ☆ ■ ✉
Rte. 1, Box 278
Blowing Rock, NC 28605
(704)264-2792
*Personal care, Companion
animal*

♥ Crebel Spa Energy
(see Cosmyl)

♥ Creighton's Naturally
(see RJG)

♥ Creme de la Terre ☆ ●
30 Cook Road
Stamford, CT 06902
(800)260-0700,
(203)324-4300
Personal care

▼ Creme Puff Pressed
Powder Compact
(see Max Factor)

♥ Creme Rouge
(see Natures Colors
Cosmetics)

▼ Crest
(see Procter & Gamble)

♥ Cross-Link Spray
(see Alpha 9)

♡ Crowne Royale Ltd. ■
P.O. Box 5238
99 Broad Street
Phillipsburg, NJ 08865
(800)992-5400,
(908)859-6488
*Cosmetics, Personal care,
Companion animal*

♥ Crystal Mist
(see Deodorant Stones
of America)

♥ Crystal Orchid
(see U.S. Sales Service)

▼ Crystal White Liquid
Detergent
(see Colgate-Palmolive)

♥ Crystalline Cosmetics, Inc. ✉
8436 North 80th Place
Scottsdale, AZ 85258-2213
(602)991-1704
Personal care

▼ Curel
(see Bausch & Lomb)

▼ Curex
(see Chesebrough-Pond's)

▼ Curl Free
(see Gillette)

♥ Curls & Color
(see Lanza Research Int'l)

▼ Curve'N Body
(see Gillette)

▼ Custom Plus
(see Gillette)

♥ CRUELTY FREE
Does not test products
or ingredients on animals

♡ Ingredients MAY
be tested on
animals

▼ Tests products or
ingredients
on animals

▼ Cut Guard
(see Chesebrough-Pond's)

▼ Cutex
(see Unilever U.S.)

❤ Cuticura
(see Dep Corporation)

❤ Daily Grind
(see Sojourner Farms
Natural Pet Products)

▼ Daisy
(see Gillette)

❤ Dallas Manufacturing
(see Color & Herbal Co.)

▼ Dash
(see Procter & Gamble)

▼ Dawn
(see Procter & Gamble)

❤ Decleor USA, Inc. ☆ ■ ●
500 West Avenue
Stamford, CT 06902
(800)722-2219,
(203)353-1771
Personal care

▼ Deep Magic
(see Gillette)

▼ Degree
(see Helene Curtis)

♡ Del Laboratories, Inc. ■ ●
565 Broad Hollow Road
Farmingdale, NY 11735
(516)293-7070
Cosmetics, Personal care

❤ Delby System ■
450 Seventh Avenue
New York, NY 10123-0811
(212)594-5036
Personal care

❤ delete-Leg Hair Removal/
Skin Smoothing System
(see Les Femmes)

❤ Delore for Nails
(see American Int'l, Ind.)

❤ Dena Corporation ☆ ■
850 Nicholas Boulevard
Elk Grove Village, IL 60007
(708)593-3041
Personal care

❤ Deneuve
(see Avon)

▼ Denquel Toothpaste
(see Procter & Gamble)

☆ Products contain no
animal or animal
derived ingredients ■ Manufacturer ● Distributor ✉ Mail Order

💜 Deodorant Stones of America (DSA) ☆ ■ ●
9420 E. Doubletree
Ranch Road, #C101
Scottsdale, AZ 85258-5508
(800)666-0373,
(602)437-8944
Personal care

💜 Dep Corporation ☆ ■
2101 East Via Arado Avenue
Rancho Dominguez, CA
90220-6189
(800)367-2855,
(310)604-0777
Personal care

▼ Depend
(see Kimberly-Clark)

💜 Derma E ■
9660 Cozycroft Avenue
Chatsworth, CA 91311
(800)521-3342,
(818)718-1420
Personal care

💜 Derma Guard
(see Pro-Tec Pet Health)

💜 Derma-Glo Skin Care
(see Nutri-Cell, Inc.)

💜 Derma-Life
Corporation ☆ ■
4149 Montgomery Blvd., N.E.
Albuquerque, NM 87109
(508)888-1789
Cosmetics, Personal care

💜 Dermalive-Skin Gel
(see Earth Solutions)

▼ Dermassage
(see Colgate-Palmolive)

💜 Dermatologic Cosmetic
Laboratories ☆ ■ ●
360 Sackett Point Road
North Haven, CT 06473
(800)552-5060,
(203)288-8633
Cosmetics, Personal care

💜 Dermatone
(see Essential Products
of America)

💜 Dermatone Lab Inc. ■ ● ✉
80 King Spring Road
Windsor Locks, CT 06096
(800)225-7546
Personal care

💜 Dermelle Botanical
Cleansing Bar
(see Essque Bodycare)

💜 CRUELTY FREE
Does not test products
or ingredients on animals

💜 Ingredients MAY
be tested on
animals

▼ Tests products or
ingredients
on animals

❤ Desert Essence ☆ ■
9510 Vassar Ave., Unit A
Chatsworth, CA 91311
(800)TI-TREE1,
(818)709-5900
Cosmetics, Personal care

❤ Desert Naturels ☆ ■
83-612 Avenue 45, Suite 5
Indio, CA 92201
(800)243-4435,
(619)775-5800
Personal care

❤ Desert Pride Yucca
(see Botanical Products)

❤ Desert Whale Jojoba Co.,
Inc. ☆ ■
P.O. Box 41594
Tucson, AZ 85717
Personal care

▼ Designer Eyes Brush &
Brow Eyebrow Color
(see Max Factor)

♡ DeSoto Inc. ■
900 East Washington St.
Joliet, IL 60434
(800)544-2814,
(815)727-4931
Household

❤ Destiny Perfume
(see Marilyn Miglin L.P.)

❤ Detox Bath Crystals
(see Wise Ways Herbals)

❤ Devi, Inc. ■ ● ✉
P.O. Box 377
Lancaster, MA 01523
(800)BEST-221,
(508)368-0066
Cosmetics, Personal care

❤ Dial Corporation ■
1850 North Central
Phoenix, AZ 85077
(602)258-3425
*Cosmetics, Personal care,
Household,
"Moratorium"*

▼ Diamond Hard Nail
Enamel
(see Max Factor)

❤ Diaper Sweet
(see Dial Corporation)

❤ Dioressence
(see Christian Dior
Perfumes)

💜 Diorissimo
(see Christian Dior Perfumes)

▼ Dippity-Do
(see Gillette)

💜 Dishmate
(see Earth Friendly
Products)

▼ Disney Toothbrushes
(see Chesebrough-Pond's)

💜 Dobie
(see Dial Corporation)

💜 Dodman's, Inc. ✉
411 White Oak Drive
Perry, MI 48872
Cosmetics, Personal care

💜 Don't Bug Me,
Inc. ☆ ■ ● ✉
326 Broad Street
Red Bank, NJ 07701
(908)933-0038
Personal care

💜 DoneGon
(see Safer Chemical)

💜 Dose Of Color
(see BeautiControl
Cosmetics)

▼ Double Edge
(see Gillette)

▼ Dove
(see Lever Brothers)

▼ DowBrands, Inc.
P.O. Box 68511
9550 N. Zionsville Road
Indianapolis, IN 46268-0511
(317)873-7000

▼ Downy
(see Procter & Gamble)

💜 DPM
(see Beauty Time)

♡ Dr. AC Daniels, Inc. ☆ ■
109 Worcester Road
Webster, MA 01570
(800)547-3760, (508)943-5563
Companion animal

💜 Dr. E.H. Bronner
(see All-One-God-Faith)

💜 Dr. Gomez Shampoo's
(see Jojoba Resources)

💜 Dr. Goodpet ■
P.O. Box 4489
Inglewood, CA 90309
(800)222-9932
Companion animal

❤ Dr. Harvey's
(see Harvey Universal)

❤ Dr. Hauschka
Cosmetics ● ✉
59C North Street
Hatfield, MA 01038
(413)247-9907
Cosmetics, Personal care

❤ Dr. Minto's
(see Minto Industries,Ltd.)

❤ Dr. Pati's
(see Auromere
Ayurvedic Imports)

❤ Dr. Singha's Mustard Bath
(see Natural Therapeu-
tics Centre)

▼ Drackett Company
(see S.C. Johnson & Son)

▼ Drain Power
(see Reckitt & Colman)

♡ Drakkar Noir
(see L'Oreal of Paris)

▼ Drano
(see S.C. Johnson & Son)

▼ Dreft
(see Procter & Gamble)

▼ Drive
(see Lever Brothers)

❤ Dry Creek Herb
Farm ☆ ■ ✉
13935 Dry Creek Road
Auburn, CA 95602
(916)878-2441
Personal care

▼ Dry Idea
(see Gillette)

▼ Dry Look
(see Gillette)

▼ Dry Style
(see DowBrands, Inc.)

❤ DSS
(see LaDove)

❤ DuBarry Products
(see Carme, Inc.)

❤ Dune
(see Christian Dior
Perfumes)

❤ Dura Green
(see Green Mountain)

☆ Products contain no   ■ Manufacturer   ● Distributor   ✉ Mail Order
animal or animal
derived ingredients

▼ Dust Guard
(see Alberto-Culver)

▼ Duster Plus
(see S.C. Johnson & Son)

💜 Dutch
(see Dial Corporation)

▼ Dynamo
(see Colgate-Palmolive)

💜 E. Burnham Company, Inc. ■
7117 N. Austin Ave.
Niles, IL 60714
(708)647-2121
Personal care

▼ Earth Born
(see Gillette)

💜 Earth Care ☆ ✉
P.O. Box 8507
Ukiah, CA 95482
(707)468-9292
Household

💜 Earth Doctor ☆ ●
828 Kings Road
Kendicott, NY 12303
Cosmetics, Personal care, Household, Companion animal

💜 Earth Enzymes
(see Earth Friendly Products)

💜 Earth Friendly Products
(see Venus Laboratories, Inc.)

💜 Earth Preserv, Ltd. ■ ● ✉
580 Decker Drive, Suite 204
Irving, TX 75062
(214)717-0399, (214)717-1669
Personal care

💜 Earth Rouge
(see Natures Color Cosmetics)

💜 Earth Science, Inc. ■
23705 Via Del Rio
Yorba Linda, CA
92687-2717
(800)222-6720, (714)692-7190
Cosmetics, Personal care

💜 Earth Sensitive Home Care
(see Espial)

💜 Earth Solutions, Inc. ☆ ■ ●
427 Moreland Ave., #100
Atlanta, GA 30307
(800)883-3376,
(404)525-6167
Cosmetics, Personal care

💜 CRUELTY FREE
Does not test products
or ingredients on animals

💜 Ingredients MAY
be tested on
animals

▼ Tests products o
ingredients
on animals

💜 Earthchild
(see Autumn Harp)

💜 Earthly Matters ☆ ■
2719 Phillips Highway
Jacksonville, FL 32207
(800)398-7503,
(904)398-1458
Household

💜 EarthRite Products
(see RCN Products, Inc.)

♡ Earthues
(see Twinscents)

▼ Easy Wash
(see Reckitt & Colman)

▼ Easy-Off
(see Reckitt & Colman)

💜 Eau Sauuage
(see Christian Dior
Perfumes)

💜 Ecco Bella
Botanicals ☆ ■
1133 Route 23
Wayne, NJ 07470
(201)696-7766
Cosmetics, Personal care

💜 Eco Source ☆ ✉
P.O. Box 1656
Sebastopol, CA 95473
(707)829-7562
*Personal care, Household,
Companion animal*

💜 Eco-Choice, Inc. ☆ ■
P.O. Box 281, Dept. 1864,
Montvale, NJ 07645-0281
(201)930-9046
*Cosmetics, Personal care,
Household, Companion
animal*

💜 Eco-Dent International,
Inc. ☆ ■ ✉
3130 Spring Street
Redwood City, CA
94063-3929
(415)364-6343
Personal care

💜 Eco-Pak Canada ■
821 Tecumseh
Pointe Claire, Canada
QC H9R 4X5
(514)421-2085
*Household, Companion
animal*

💜 Eco-Zone
(see Eco-Safe Products)

☆ Products contain no   ■ Manufacturer   ● Distributor   ✉ Mail Order
animal or animal
derived ingredients

💜 Ecocare
(see Paul Mazzotta, Inc.)

💜 EcoSpun
(see Alpen Limited)

💜 Ecotopia ☆ ✉
1268 E. Honeymoon Bay Rd.
Greenbank, WA 98253
(206)331-8153
Personal care

💜 Ecover, Inc. ■
P.O. Box SS
Philmont, NY 12565
(518)672-0190
Household

💜 Eden Botanicals ☆ ■ ● ✉
15994 Stetson Road
Los Gatos, CA 95030
(408)353-8749
Personal care

💜 Eden Essence Oil
(see Eden Botanicals)

▼ Edge Shave
(see S.C. Johnson & Son)

💜 EFP Cream Polish
(see Earth Friendly
Products)

💜 EFP Toilet Bowl Cleaner
(see Earth Friendly
Products)

💜 Egyptian Earth
(see Colora)

💜 Elephant Ear
(see Delby System)

▼ Elizabeth Arden, Inc.
1345 Avenue of the Americas
New York, NY 10105
(212)261-1000

💜 Ella Bache, Inc. ■
8 West 36th Street
New York, NY 10018
(212)279-0842
Personal care

💜 Ellegance
(see Naturelle Cosmetics)

💜 Ellon USA, Inc. ☆ ■ ● ✉
644 Merrick Road
Lynbrook, NY 11563
(800)4BE-CALM
*Personal care, House-
hold, Companion animals*

♥ Elysee Scientific
Cosmetics ■ ●
6804 Seybold Road
Madison, WI 53719
(608)271-3664
Cosmetics, Personal care

♥ EM Enterprises ☆ ✉
41964 Wilcox Rd., Dept. AV
Hat Creek, CA 96040
(916)335-7097
Personal care, Household

♥ Emerald Forest
(see Natural Nectar)

▼ Endust
(see Kiwi Brands)

♥ Energizer Treatment
Shampoo
(see Hobe Labs)

♡ English Leather
(see Mem Company)

♥ Entourage/Biogime ■ ●
1665 Townhurst, #100
Houston, TX 77043
(800)338-8784,
(713)827-1972
Household, Personal care

▼ Envy
(see S.C. Johnson & Son)

♥ Epilady International ☆ ■ ●
39 Cindy Lane
Ocean, NJ 07712
(800)879-LADY,
(903)493-3232
Personal care

▼ Era
(see Procter & Gamble)

▼ Erase Cover-up
(see Max Factor)

▼ Erno Laszlo
200 First Stamford Place
Stamford, CT 06902-6759
(203)462-5700

♥ Es-Gen
(see Professional &
Technical Services)

▼ Esoterica
(see SmithKline Beecham)

♥ Espial, Inc. ☆ ■ ●
7045 S. Fulton St., #200
Englewood, CO 80112-3700
(303)799-0707
*Cosmetics, Personal care,
Household*

💜 Espree Animal
Products, Inc. ☆ ■
6015 Commerce Drive
Suite 400
Irving, TX 75063-2665
(800)328-1317
Companion animal

💜 Espree Associates, Inc. ☆ ■
13581 Pond Springs Rd.
Suite 308
Austin, TX 78729-4425
(512)263-9184
Cosmetics

💜 Espree Organic
Shampoo ☆ ■ ●
2921 Suffolk Ct., East
Suite 460
Fort Worth, TX 76133
(214)756-0626
Companion animal

💜 Essee Salon & Spa
(see Catalog Ventures)

💜 Essensual Moments
(see Pacific Scents)

💜 Essentia
(see Shirley Price
Aromatherapy)

💜 Essential Aromatics ☆ ■
205 N. Signal St.
Ojai, CA 93023
(805)640-1300
Cosmetics, Personal care

💜 Essential Elements ☆ ■ ●
2415 Third Street, #235
San Francisco, CA 94107
(415)621-9881
Personal care

💜 Essential Products of
America, Inc. ☆ ■
5018 North Hubert Ave.
Tampa, FL 33614
(800)822-9698, (813)877-9698
Cosmetics, Personal care

💜 Essque Bodycare ■
Box 7635
Greenwich, CT 06836
(203)869-9658
Personal care

💜 Estee Lauder Companies ■
767 Fifth Avenue
New York, NY 10153
(212)572-4200
Cosmetics, Personal care

▼ ETC
(see DowBrands, Inc.)

💜 CRUELTY FREE
Does not test products
or ingredients on animals

💜 Ingredients MAY
be tested on
animals

▼ Tests products or
ingredients
on animals

❤ European Gold ☆ ■ ●
33 SE 11th Street
Grand Rapids, MN 55744
(612)326-0266,
(800)232-4518
Personal care

❤ European Soaps, Ltd. ●
12300 15th Ave. NE
Seattle, WA 98125
(206)361-9143
Personal care

❤ Eva Jon Cosmetics ✉
1016 E. California St.
Gainesville, TX 76240
(817)668-7707
Cosmetics, Personal care,
Companion animal

❤ Ever Young, Inc. ☆ ●
55 W. Sunset Way
Ilssaquah, WA 98027
(206)391-1584
Cosmetics, Personal care

♡ Everclean
(see Home Health Products)

❤ Eversoft
(see Andrew Jergens)

❤ Everybody Ltd. ■ ✉
1175 Walnut Street
Boulder, CO 80302-5116
(800)748-5675
Cosmetics, Personal care,
Household

♡ Evolution
(see Guerlain, Inc.)

❤ Exotic Nature ■
P.O. Box 834
Cambria, CA 93428
(805)927-2517
Personal care

❤ Expressions Hair Products
(see IQ Products)

❤ Exquissite Plus
(see Wachters')

❤ Extra Save
(see Associated
Wholesale Grocers)

▼ Eye Make-up Remover
(see Max Factor)

♡ FA Cosmetics Imports
International Corp. ☆ ●
1141 Westwood Blvd.
Los Angeles, CA 90024
(310)208-4558
Personal care

▼ Fab
(see Colgate-Palmolive)

▼ Faberge
(see Unilever U.S.)

▼ Fabuloso
(see Colgate-Palmolive)

❤ Face Feminizer
(see BeautiControl
Cosmetics)

▼ Face Saver
(see Gillette)

❤ Face to Face ■ ●
18399 Ventura Blvd., #10
Tarzana, CA 91356-4233
Cosmetics, Personal care

♡ Face Up
(see AHC Pharmacal)

❤ Facets
(see Crystalline Cosmetics)

❤ Fahrenheit
(see Christian Dior
Perfumes)

❤ Faith In Nature
(see Faith Products)

❤ Faith Products, Ltd. ☆ ■
5 Kay Street
Bury, Lancashire BL9 6BU
England UK
1-61-764-2555
*Cosmetics, Household,
Personal care*

❤ Family Essentials
(see Pacific Scents)

▼ Fantastik
(see DowBrands)

▼ Fasteeth Denture Adhesive
(see Procter & Gamble)

♡ Fathom
(see Mem Company)

❤ Fautless Starch/Bon
Ami Company ■
1025 West 8th Street
Kansas City, MO 64101
(816)842-1230
Household

▼ Favor
(see S.C. Johnson & Son)

▼ FDS
(see Alberta-Culver)

▼ Featherblend Kohliner
(see Max Factor)

♥ Fels Naptha
(see Dial Corporation)

▼ Fem Mist
(see DowBrands, Inc.)

♥ Fermodyl Professionals, Inc.
(see Revlon)

♥ Fiber-off
(see Alpha 9)

▼ Final Net
(see Clairol)

▼ Final Touch
(see Lever Brothers)

▼ Finale
(see Clairol)

♥ Finelle Cosmetics ☆ ■
137 Marston Street
Lawrence, MA 01842-2808
(508)682-6112
Cosmetics, Personal care

▼ Finesse
(see Helene Curtis)

♥ Finley Pharmaceuticals ☆ ■
7046 Summit Drive
Navarre, FL 32566-8745
(602)722-3939
Personal care

♥ Finnfoods ☆ ■
500 N. Field Drive
Lake Forest, IL 60045
(708)735-7819
Personal care

♥ Firm & Fill
(see Palm Beach Beauty Products)

▼ Fisherman's Friend
(see Bristol-Myers Squibb)

▼ Fixodent
(see Procter & Gamble)

▼ Flair
(see Gillette)

♡ Flame Glow ■ ●
565 Broad Hollow Road
Farmingdale, NY 11735
(516)293-7070
Cosmetics

♥ Flare Cologne
(see BeautiControl
Cosmetics)

💜 Flea Pruff For Carpets
(see Copper-Brite)

💜 Fleabusters
(see Rx for Fleas, Inc.)

💜 Flex
(see Revlon)

💜 Flower Essence
Services ☆ ■ ● ✉
P.O. Box 1769
Nevada City, CA 95959
(800)548-0075
Cosmetics, Personal care

💜 Fluff
(see Austin's)

▼ Foamy Shaving Cream
(see Gillette)

💜 Focus 21 International,
Inc. ■
Oakridge Business Ctr.
2755 Dos Aarons Way
Vista, CA 92083
(800)821-9401, (619)727-6626
Personal care, Household

▼ Foot Guard
(see Gillette)

💜 Footherapy
(see Para Labs)

▼ For Oily Hair Only
(see Gillette)

💜 Forest Essentials ☆ ■
1718 22nd Street
Santa Monica, CA 90404
(310)264-1997
Personal care

💜 Forest Pure
(see Levlad, Inc.)

💜 Forever 29
(see Palm Beach Beauty Products)

💜 Forever New Int'l, Inc. ☆ ■
4701 N. Fourth Avenue
Sioux Falls, SD 57104-0403
(800)456-0107,
(605)331-2910
Household

▼ Formula 409
(see Clorox)

💜 Fort Howard Corporation ☆ ■
P.O. Box 19130
Green Bay, WI 54307-9130
(414)435-8821
Household

▼ Fostex
(see Bristol-Myers Squibb)

❤ 4 The Planet
(see Zenith Advanced
Health Systems Int'l, Inc.)

❤ Four (IV) Trail
Products ☆ ■ ● ✉
P.O. Box 1033
Sykesville, MD 21784
(410)795-8989
Companion animal

♡ Fragrance Puffs
(see Botanicus)

❤ Fragrant Essence
(see Blessed Herbs)

❤ Framesi USA/Roffler ☆ ■
400 Chess Street
Coraopolis, PA 15108
(412)269-2950
Cosmetics

❤ Francosmetic Int'l,
Inc. ☆ ■ ● ✉
8601 Wilshire Blvd., #604
Beverly Hills, CA 90211
(310)659-1970
Cosmetics, Personal care

❤ Frank T. Ross & Sons, Ltd. ☆ ■
6550 Lawrence Avenue
East Scarborough
Ontario, Canada M1C 4A7
(416)282-1107
Personal care, Household

♡ Fred Hayman Beverly
Hills, Inc. ■ ●
190 Canon Drive, Suite 400
Beverly Hills, CA
90210-5315
(310)271-3100
Cosmetics

❤ Free Form
(see Minto Ind., Ltd.)

❤ Free Spirit Enterprises ☆ ■
P.O. Box 2638
Guerneville, CA 95446
(707) 869-1942
Personal care

❤ Freeman Cosmetics
Corporation ■ ✉
P.O. Box 4074
Beverly Hills, CA 90213
(800)FREEMAN,
(310)470-6840
Personal care

☆ Products contain no   ■ Manufacturer ● Distributor ✉ Mail Order
animal or animal
derived ingredients

❤ French Transit, Ltd. ☆ ■ ●
398 Beach Road
Burlingame, CA 94010
(800)829-ROCK,
(415)548-9600
Cosmetics, Personal care

❤ Fresh Foot
(see Deodorant Stones of
America)

♡ Fresh Fruit Soaps
(see Botanicus)

▼ Fresh Start
(see Colgate-Palmolive)

▼ Fresh Step Catbox Filler
(see Clorox)

❤ Friendly Systems Inc. ☆ ■
2727 Chemsearch
Irving, TX 75062
(800)535-3066
Household

♡ Frontier Cooperative Herbs
(see Aura Cacia)

❤ Frontier Products
(see Carme)

▼ Frost & Tip
(see Clairol)

❤ Fruit of the Earth, Inc. ■ ●
P.O. Box 152044
Irving, TX 75015-2044
(800)527-7731, (214)790-0808
Cosmetics, Personal care

❤ Fuller Brush
Company ■ ● ✉
P.O. Box 1247
Great Bend, KS 67530
(316)792-1711
Personal care, Household

▼ Future
(see S.C. Johnson & Son)

❤ Futurebiotics ■
72 Cotton Mill Hill
Brattleboro, VT 05301
Personal care

❤ Gaea
(see Greenway Products)

▼ Gain
(see Procter & Gamble)

❤ Gajee Herbal Ayurvedic
Cosmetics ■ ✉
1661 Botehlo Drive
Suite 290, Dept E
Walnut Creek, CA 94596
Cosmetics, Personal care

❤ CRUELTY FREE
Does not test products
or ingredients on animals

❤ Ingredients MAY
be tested on
animals

▼ Tests products or
ingredients
on animals

❤ Gannons ☆ ■ ✉
1020 7th Street
Portsmouth, OH 45662-4105
(614)353-1667
Household, *Companion
animal*

❤ Garcoa Labs ☆ ■
6324 Variel, Ste. 301
Woodland Hills, CA 91367
(818)887-3705
Personal care

❤ Garden Botanika ■
8624 154th Ave., N.E.
Redmond, WA 98052
(800)968-7842, (206)881-9603
Cosmetics, Personal care

❤ Garden Bouquet
(see Dial Corporation)

❤ Gee, Your Hair Smells
Terrific
(see Andrew Jergens)

❤ Gem
(see Delby System)

❤ Gena Labs ■
P.O. Box 380459
Duncanville, TX 75138
(214)296-2887
Cosmetics, Personal care

❤ Gena Shampoos
(see Gena Labs)

❤ General Nutrition
Corporation ■ ●
921 Penn Avenue
Pittsburgh, PA 15222
(800)477-4462,
(412)288-2042
Cosmetics, Personal care

❤ Genki USA ☆ ●
P.O. Box 56126
Los Angeles, CA 90008
Personal care

❤ Gentle Breezes
Tub-N-Tile Cleaner
(see Earthly Matters)

❤ Georgette Klinger
Inc. ■ ● ✉
501 Madison Avenue
New York, NY 10022
(800)KLINGER,
(212)838-3200
Cosmetics, Personal care

♡ Georgia-Pacific
Corporation ☆ ■ ●
133 Peachtree St., NE
Atlanta, GA 30303
Household

☆ Products contain no
animal or animal
derived ingredients   ■ Manufacturer   ● Distributor   ✉ Mail Order

💜 Gerda Spillman Swiss
Skin Care
(see Mar-Riche Ent.)

💜 Geremy Rose Fresh
(see New Moon Extracts)

💜 Germaine Monteil
Cosmetiques
(see Revlon)

💜 Gift of Life
(see Para Labs)

▼ Gillette Company
Prudential Tower Building
Boston, MA 02199
(617)463-3000

💜 Ginza
(see Conair)

♡ Giorgio Armani
(see L'Oreal of Paris)

▼ Giorgio Beverly Hills
(see Procter & Gamble)

♡ Giovanni Hair Care
Products ☆ ■
P.O. Box 39378
Los Angeles, CA 90039
(213)563-0355
Personal care

💜 Glad Rags ☆ ■ ● ✉
P.O. Box 12751
Portland, OR 97212
Personal care

▼ Glade
(see S.C. Johnson & Son)

▼ Glamorene Carpet Cleaner
(see Reckitt & Colman)

💜 Glass Mate
(see Green Mountain)

▼ Glass Plus
(see DowBrands)

▼ Glass Works
(see Miles)

▼ Gleem
(see Procter & Gamble)

▼ Glints
(see Clairol)

💜 Glo Marr Products
(see Kenic Pet Products)

▼ Glo-Coat
(see S.C. Johnson & Son)

♡ Gloria Vanderbilt
(see L'Oreal of Paris)

💜 CRUELTY FREE
Does not test products
or ingredients on animals

💜 Ingredients MAY
be tested on
animals

▼ Tests products or
ingredients
on animals

▼ Glory
(see S.C. Johnson)

♡ Glover Hair Products ■
1645 Oak Street
Lakewood, NJ 08701
(908)905-5252
Personal care

♥ Gly-Miracle
(see Palm Beach Beauty
Products)

♡ Gold Shield ●
P.O. Box 858
Mahwah, NJ 07430
(201)529-4900
Personal care

♥ Golden Harvest
(see General Nutrition)

♥ Golden Lotus
(see Green Mountain)

♥ Golden Pride - Rawleigh ☆ ●
1501 Northpoint Parkway
Suite 100
West Palm Beach, FL 33407
(407)640-5700
Personal care, Household

♥ Golden Star, Inc. ☆ ■
400 E. 10th Avenue
North Kansas City, MO
64116
(816)842-0233
Household

♥ Goldwell ■ ●
9050 Junction Dr
Annapolis Junction, MD
20701
(800)288-9118,
(301)725-6620
Cosmetics, Personal care

♥ Good Clean Fun
(see Smith & Vandiver)

▼ Good Measure
(see S.C. Johnson & Son)

▼ Good News
(see Gillette)

▼ Good Sense
(see S.C. Johnson & Son)

♥ Goodebodies USA,
Inc. ☆ ■ ●
1001 S. Bayshore Dr.,
#2402
Miami, FL 33131
(305)358-1903
Cosmetics, Personal care

💜 Goodier, Inc. ☆ ■
9027 Diplomacy Row
Dallas, TX 75247
(214)630-1803
Personal care

💜 Grace Cosmetics
(see Pro-Ma Systems)

💜 Grande Finale
(see Conair)

▼ Graneodin
(see Bristol-Myers Squibb)

💜 Granny's Int'l, Inc. ■ ●
P.O. Box 218
Karlstad, MN 56732
Personal care, Household

💜 Granny's Old Fashioned
Products ☆ ■
P.O.Box 660037
Arcadia, CA 91066
Personal care, Household

▼ Great Lady
(see Unilever U.S.)

💜 Great Mother's
Goods ☆ ■ ● ✉
585 E. 31st Street
Durango, CO 81301
(970)247-4687
Personal care

💜 Great Stuff
(see Hobe Labs)

💜 Green Ban ☆ ■
Box 146
Norway, IA 52318
(319)446-7495
*Personal care, Companion
animal*

💜 Green Mountain
Products, Inc. ☆ ■
517 Memphis Junction Rd.
Bowling Green, KY 42102
(502)796-8353
*Cosmetics, Personal care,
Household*

💜 Green Thoughts Mildew
Stain Remover
(see Earthly Matters)

💜 Greener Pastures
Multipurpose Cleaner
(see Earthly Matters)

♥ Greenhouse Ventures ■ ✉
5121 Wagner Way
Agoura, CA 91301-4731
Personal care

♥ Greenway Products, Inc. ☆ ✉
P.O. Box 183
Port Townsend, WA 98368
(800)966-1445,
(206)385-7124
Personal care, Household,
Companion animal

▼ Groom & Clean
(see Chesebrough-Pond's)

♥ Gruene Skin Care Products
(see Human Kind)

♥ Gryphon Development
Inc. ●
767 5th Avenue
New York, NY 10153
(800)395-1001,
(212)527-7382
Cosmetics, Personal care

♡ Guerlain, Inc. ☆ ●
444 Madison Avenue
17th Flr
New York, NY 10022
(212)751-1870
Cosmetics

♡ Guy Laroche
(see L'Oreal of Paris)

♡ h.e.r.c. Consumer
Products, L.L.C. ■
1 Cummings Point Road
Stamford, CT 06904
(203)351-9000
Household

♥ H2O Plus, Inc. ■ ●
676 N. Michigan Ave.
39th Floor
Chicago, IL 60611
(800)242-BATH,
(312)642-1100
Cosmetics, Personal care

♥ Habanita Fragrance &
Bath Lines
(see Dep Corporation)

♥ Hair Doc Company ●
3139 Los Feliz Drive
Thousand Oaks, CA 91362
(805)373-1637
Personal care

♥ Hair Lover's
(see Hobe Labs)

♥ Hair Management for Men
(see Conair)

☆ Products contain no   ■ Manufacturer   ● Distributor   ✉ Mail Order
animal or animal
derived ingredients

💜 Hair Saver
(see Palm Beach Beauty Products)

💜 Hair Therapy
(see Palm Beach Beauty Products)

💜 Halsa Swedish Botanical
(see Dep Corporation)

💜 Halston
(see Revlon)

▼ Handi Wipes
(see Colgate-Palmolive)

💜 Handy Pantry ☆ ■ ●
1520 East Cedar St.
Tempe, AZ 85281
Personal care

💜 Hansen's Pet Products
Company ☆ ■
4935 Warner Ave.
Huntington Beach, CA
92648-5181
(714)846-7375
Companion animal

▼ Happy Face
(see Gillette)

💜 Hargen Distributors,
Inc. ■ ●
3422 W. Wilshire Drive, #13
Phoenix, AZ 85009-1457
(602)278-6046
Personal care

💜 Harvey Universal, Inc. ☆ ■
1805 W. 208th Street
Torrance, CA 90501
(310)328-9000,
(800)800-3330
Household

💜 Harvey's Rug Shampoo
Concentrate
(see Harvey Universal)

💜 Hawaiian Face
(see Free Spirit Ent.)

💜 Hawaiian Fruit Enzyme
Exfoliant
(see Alexander Avery
Purely Natural Skin Care)

💜 Hawaiian Resources Co.,
Ltd. ☆ ●
94527 Puahi Street
Waipahu, HI 96797
Cosmetics, Personal care

▼ Hawk After Shave
(see Mennen)

♡ Heaven Scent
(see Mem Company)

▼ Head & Shoulders
(see Procter & Gamble)

♥ Heavena
(see Avanza)

♥ Head Shampoo, Inc./
Pure & Basic Products ☆ ■
20625 Belshaw Avenue
Carson, CA 90746
(800)432-3787,
(310)868-1630
Personal care, Household

♥ Helen Lee Skin Care &
Cos. ● ✉
205 E. 60th St.
New York, NY 10022
(800)288-1077,
(212)888-1233
Cosmetics, Personal care

▼ Heads Up
(see Gillette)

▼ Helene Curtis Industries, Inc.
325 North Wells Street
Chicago, IL 60610-4791
(312)661-0222

♥ Healing Herbs Flower
Essences
(see Flower Essence Services)

▼ Helene Rubenstein
(see Colgate-Palmolive)

♥ Health From The Sun ☆ ● ✉
P.O. Box 840
Sunapee, NH 03782
Cosmetics, Personal care

♥ Henri Bendel
(see Gryphon Development)

♥ Herb Garden ☆ ■ ✉
P.O. Box 773-N
Pilot Mountain, NC 27041
(910)368-2723
*Personal care, Household,
Companion animal*

♥ Healthy Times ☆ ■
461 Vernon Way
El Cajon, CA 92020
(619)593-2229
Personal care

☆ Products contain no
animal or animal
derived ingredients
■ Manufacturer ● Distributor ✉ Mail Order

▼ Herbal Essences
(see Clairol)

💜 Herbal Melange
(see Norimoor)

💜 Herbal Products &
Development ☆ ■ ✉
P.O. Box 1084
Aptos, CA 95001
(408)688-8706
Personal care, Household

💜 Herbal Savvy
(see Country Comfort)

💜 Herbal Soft Lotion
(see Gena Labs)

💜 Herbal Works
(see New Age Creations/
Jeanne Rose)

💜 Herbomineral
(see AuromereAyurvedic
Imports)

💜 Heritage Store, Inc. ■ ✉ ●
P.O. Box 444
Virginia Beach, VA 23458
(804)428-0100
*Cosmetics, Personal care,
Household*

▼ hero
(see Chesebrough-Pond's)

▼ Hi-Dri
(see Kimberly-Clark)

💜 Hilex
(see Dial Corporation)

▼ Hill's Pet Products
(see Colgate-Palmolive)

💜 Hobe Labs, Inc. ☆ ■
4032 E. Broadway Road
Phoenix, AZ 85040
(800)528-4482
Personal care

▼ Hold & Hold & Hold
(see DowBrands, Inc.)

💜 Holloway House Inc. ☆ ■ ●
P.O. Box 50126
8328 Masters Road
Indianapolis, IN 46250
(800)255-1891
Household

♡ Home Health Products
Company ■ ✉
P.O. Box 3130
Virginia Beach, VA 23454
(804)468-3130
Cosmetics, Personal care

💜 CRUELTY FREE
Does not test products
or ingredients on animals

♡ Ingredients MAY
be tested on
animals

▼ Tests products or
ingredients
on animals

▼ Home Perms
(see Gillette)

💜 Home Service Products
Company ☆ ■ ● ✉
P.O. Box 245
Pittstown, NJ 08867
(908)735-5988
Household

💜 Homebody/Perfumoils,
Inc. ●
P.O. Box 2266
W. Brattleboro,VT
05303-2266
(802)254-6280
Cosmetics, Personal care

♡ Homesteader's Pumice
Compound
(see Homesteader's Soap)

♡ Homesteader's Soap Co. ■
P.O. Box 66
Middle Falls, NY 12848
(518)692-9469
Personal care

💜 Honey Silk
(see Beehive Botanicals)

💜 Horseman's Dream ■ ●
P.O. Box 26797
Fort Worth, TX 76126
(817)560-8818
Companion animal

▼ Hot One Shave Creams
(see Gillette)

♡ House of Lowell, Inc. ☆ ■
10695 Stargate Lane
Cincinnati, OH 45240
(513)851-8514
Cosmetics, Personal care

▼ Huggies
(see Kimberly-Clark)

▼ Hugo Boss
(see Procter & Gamble)

💜 Huish Detergents,
Inc. ☆ ■ ●
3540 West 1987 South
Salt Lake City, UT 84104
(800)776-6702, (801)975-3100
Household

💜 Human Kind ☆ ■ ●
8530 Venice Blvd.
Los Angeles, CA 90034
Personal care

♥ Hund-N-Flocken Dog
Food Flakes
(see Solid Gold Holistic)

♥ Hydrating Germanium
Creme
(see Wachters')

♥ Hygenic Cosmetics, Inc. ☆ ■
5-02 Banta Place
Fairlawn, NJ 07410
Cosmetics, Personal care

♡ Hysan Corporation ☆ ■
2929 Allen Pkwy., Ste 2130
Houston,TX 77019
(713)620-7700
Household

♥ I Shawn
(see La Dove)

♥ I-Tech Laboratories ☆ ●
P.O. Box 9
Jericho, NY 11753
(516)938-0088
Cosmetics, Personal care

♥ Ida Grae Cosmetics
(see Nature's Colors
Cosmetics, Ltd.)

♥ Ideal Shape
(see Cernitin America)

♡ Il-Makiage Inc. ■ ● ✉
45-49 Davis Street
Long Island City, NY 11101
(800)722-1011,
(718)361-3123
Cosmetics, Personal care

♥ Ilona ■ ● ✉
629 Park Avenue
New York, NY 10021
(800)622-4355
Cosmetics, Personal care

▼ Image Body Spray
(see Gillette)

♥ Image Labs, Inc. ☆ ■
2340 Eastman Avenue
Oxnard, CA 93030
(800)421-8528
Personal care

♥ Imari Fragrance
(see Avon Products)

♥ Impressions
(see Melaleuca)

♥ Impressions Ultimate
Shaper Spray
(see IQ Products)

▼ Impulse
(see Chesebrough-Pond's)

▼ Incognito
(see Procter & Gamble)

▼ Incredible Blue Mask
(see Max Factor)

♡ Indian Creek Naturals ■ ✉
P.O. Box 63
591 Indian Creek Rd.
Selma, OR 97538
(503)592-2616
Personal care

♥ Infinite Quality, Inc. ☆ ■
8245 S. King Dr.
Chicago, IL 60619
(312)783-4900
Personal care

▼ Infusion 23
(see Clairol)

▼ Instant Hair Set & Style
(see DowBrands, Inc.)

♥ Institute of Trichology ■
1619 Reed St.
Lakewood, CO 80215
(800)458-8874,
(303)232-6149
Personal care

♥ Internatural ☆ ✉
P.O. Box 1008
1100 Lotus Drive
Silver Lake, WI 53170
Cosmetics, Personal care

▼ Invisible Make-up
Foundation
(see Max Factor)

▼ Ipana Toothpaste
(see DowBrands, Inc.)

♥ IQ Products Company ☆ ■
16212 State Highway 249
Houston, TX 77086
(713)444-6454
Personal care, Household

▼ Irish Spring
(see Colgate-Palmolive)

♡ Issima
(see Guerlain, Inc.)

♥ It
(see Key Distributors)

▼ Ivory
(see Procter & Gamble)

☆ Products contain no
animal or animal
derived ingredients

■ Manufacturer　● Distributor　✉ Mail Order

♥ Izy's Skin Care
Products ☆ ■ ●
13399 Terry
Detroit, MI 48227
(313)836-2675
Cosmetics, Personal care

♥ J & J Jojoba/California
Gold Products ☆ ■
7826 Timm Road
Vacaville, CA 95688
(707)447-1207
Cosmetics, Personal care

♡ J.F. Lazartigue ■ ●
764 Madison Avenue
New York, NY 10021
(212)288-2250
Cosmetics

♥ Jacki's Magic Lotion ■ ✉
258 A Street, #7-A
Ashland, OR 97520
(800)729-8428, (503)488-1388
Personal care

▼ Jaclyn Smith's California
(see Max Factor)

♥ Jafra Cosmetics, Inc. ■
P.O. Box 5026
West Lake Village, CA 91359
(800)551-2345, (805)496-1911
Cosmetics, Personal care

♥ Jamieson
(see C.E. Jamieson & Co., Ltd.)

♥ Janca's Jojoba Oil &
Seed Company ■ ● ✉
456 E. Juanita, #7
Mesa, AZ 85204
(602)497-9494
*Cosmetics, Personal care,
Companion animal*

♥ Janet Sartin Cosmetics ☆ ● ✉
500 Park Avenue
New York, NY 10022
(212)751-5858
Cosmetics

♥ Janie, The Dry Stick
Spot Remover
(see Card Corp.)

♥ Janta International
Co. (J. I. C.) ☆ ●
333 Kearney St., Ste #203
San Francisco, CA 94108
(415)591-9465
Personal care

♡ Jardin de Bagatelle
(see Guerlain, Inc.)

💜 Jason Natural
Cosmetics ☆ ■
8468 Warner Drive
Culver City, CA 90232
(800)527-6605, (310)838-7543
Personal care

▼ Javex
(see Colgate-Palmolive)

▼ Jazzing
(see Clairol)

♡ JC Garet, Inc. ☆ ■
2471 Coral Street
Vista, CA 92083
(619)598-0505
Household

💜 Jean Nate
(see Revlon)

💜 Jean-Louis Scherrer Fragrance
(see Dep Corporation)

💜 Jeanne Gatineau
(see Revlon)

💜 Jelene, Inc. ■
1332 Anderson Rd., Ste. 115
Clawson, MI 48017-1094
Personal care

💜 Jelmar Company ☆ ■ ●
6600 N. Lincoln Ave.
Lincolnwood, IL 60645
(708)675-8400
Household

💜 Jennico, Inc. ☆ ■
4404 Anderson Drive
Eau Claire, WI 54703
Household

💜 Jergens Products
(see Andrew-Jergens)

💜 Jessica McClintock ☆ ■
1400 Sixteenth Street
San Francisco, CA 94103
(800)333-5301,
(415)495-3030
Cosmetics

💜 Jheri Redding Products
(see Conair)

▼ Jhirmack
(see Playtex Family
Beauty Care)

💜 JHL
(see Estee Lauder)

💜 Jock's Rock
(see Deodorant Stones
of America)

☆ Products contain no
animal or animal
derived ingredients ■ Manufacturer ● Distributor ⊠ Mail Order

💚 Joder Sales Co.
(see I-Tech Labs)

💚 Joe Blasco
Cosmetics ☆ ■
7340 Greenbriar Pkwy.
Orlando, FL 32819
(800)553-1520,
(407)363-7070
Cosmetics, Personal care

💚 John Paul Mitchell
Systems ☆ ■
26455 Golden Valley Road
Saugus, CA 91350
(805)298-0400 (707)823-0503
Personal care

▼ Johnson & Johnson
One Johnson & Johnson
Plaza
New Brunswick, NJ 08933
(800)526-3967,
in NJ (800)942-7764

▼ Johnson's Paste Wax
(see S.C. Johnson & Son)

💚 JOICO Labs, Inc. ☆ ■
345 Baldwin Park Blvd.
City of Industry, CA 91746
Outside CA (800)445-6426,
(818)968-6111
Personal care

💚 Jojoba Farms
(see Carme)

💚 Jojoba Resources, Inc. ☆ ■ ●
6505 West Frye Road, Suite 23
Chandler, AZ 85226-3330
(800)528-4284,
(602)961-1150
Cosmetics, Personal care

💚 Jolen, Inc. ☆ ■
25 Walls Drive
Fairfield, CT 06430
(203)259-8779
Cosmetics

💚 Jones Medical Industries
(see Bronson Pharmaceuticals)

💚 Jontue
(see Revlon)

💚 Jordan
(see Dep Corporation)

💚 Joshua Solution ☆ ■ ● ⊠
4151 N. 32nd Street
Phoenix, AZ 85018
(800)992-0125,
(602)956-1560
Personal care

▼ Joy Dishwashing Liquid
(see Procter & Gamble)

💚 CRUELTY FREE
Does not test products
or ingredients on animals

💚 Ingredients MAY
be tested on
animals

▼ Tests products or
ingredients
on animals

♡ JR Carlson Laboratories,
Inc. ■
15 College
Arlington Heights, IL
60004-1985
(708)255-1600,
Outside IL (800)323-4141
Cosmetics, Personal care

♥ JR Liggett Ltd. ☆ ■
RR 2, Box 911
Cornish, NH 03745
(603)675-2055
Personal care

▼ Jubilee
(see S.C. Johnson & Son)

▼ Just Whistle
(see Gillette)

♥ Just Wonderful Hair
Care Products
(see Conair)

♡ Just 'n Time Spot Remover
(see SerVaas Labs)

♥ Juvenesse by Elaine
Gayle ☆ ■ ●
680 North Lake Shore Drive
Chicago, IL 60611
(312)944-1211
Cosmetics, Personal care

▼ Kaleidocolors
(see Clairol)

♥ Kallima Int'l ☆ ■ ● ⊠
P.O. Box 475695
Garland, TX 75047-5695
(800)KALLIMA
(214) 240-1957
Cosmetics, Personal care

♥ Kama Sutra
Company ☆ ■ ● ⊠
2260 Towngate Road
Village, CA 91361
(805)495-7479
Personal care

♥ Katonah Scentral ☆
51 Katonah Avenue
Katonah, NY 10536
(800)29-SCENT,
(914)232-7519
*Cosmetics, Personal care,
Household*

♥ Katz-N-Flocken
(see Solid Gold Holistic)

♥ Keep
(see Action Labs)

♥ Keep America Clean
(see Abkit)

💜 Kelemata Group ▪ ●
555 Madison Avenue
New York, NY 10022
(212)750-1111
Cosmetics, Personal care

▼ Kendall-Futuro Company
5405 DuPont Circle,
Suite A
Cincinnati, OH
45150-2735

💜 Kenic Pet Products, Inc. ☆ ▪
109 S. Main St.
Lawrenceburg, KY 40342
(800)228-7387,
(502)839-6996
Companion animal

💜 Kenra Laboratories, Inc. ▪
6501 Julian Avenue
Indianapolis, IN
46219-6695
(800)428-8073,
(317)356-6491
Personal care

💜 Keren Happuch,
Ltd. ☆ ● ✉
P.O. Box 809
Oconomowoc, WI 53066
(414)567-8510
Cosmetics, Personal care

▼ Keri
(see Bristol-Myers Squibb)

♡ Kettle Care ▪ ✉
710 Trap Road
Columbia Falls, MT
59912-9223
(406)756-3485
Personal care

💜 Key Distributors, Inc. ▪ ●
16035 E. Arrow Highway
Irwindale, CA 91706
(818)337-5200
Personal care

💜 Key Velvet
(see Delby System)

💜 Key West Fragrance &
Cosmetic Factory, Inc. ▪ ● ✉
P.O. Box 1079
Key West, FL 33040-1079
(800)445-ALOE,
(305)294-5592
Cosmetics, Personal care

💜 Khepra Skin Care, Inc. ☆ ▪
3939 IDS Center
80 S. 8th Street
Minneapolis, MN 55402
Personal care

♡ Kiehl's ■ ✉ ●
109 3rd Ave.
New York, NY 10003
Cosmetics

▼ Kimberly-Clark Corporation
401 North Lake Street
Neenah, WI 54957
(800)544-1847

▼ Kirk's
(see Procter & Gamble)

▼ Kirkman Borax
(see Colgate-Palmolive)

♥ Kiss My Face Corp. ☆ ■
144 Main St.
Gardiner, NY 12525-0224
(800)262-KISS,
(914)255-0884
Personal care, Cosmetics

♥ KIT Products ☆ ✉
2545-D Prairie Road
Eugene, OR 97405
(800)359-2940
Personal care

▼ Kiwi Brands Inc.
447 Old Swede Road
Douglassville, PA
19518-1239
(610)385-3041

▼ KL Fragrance
(see Unilever U.S.)

♥ Klaire Laboratories,
Inc. ■ ●
1573 Seminole
San Marcos, CA 92069
Cosmetics, Personal care

▼ Klean 'n Shine
(see S.C. Johnson & Son)

▼ Klear
(see S.C. Johnson & Son)

♡ Kleen Brite Labs, Inc. ☆ ■
P.O. Box 20408
Rochester, NY 14602-0408
(716)637-0630
Household

♥ Kleen Cleaner
(see Green Mountain)

▼ Kleen Guard
(see Alberto-Culver)

▼ Kleenex
(see Kimberly-Clark)

▼ Kleenite
(see Procter & Gamble)

☆ Products contain no
animal or animal
derived ingredients
  ■ Manufacturer ● Distributor ✉ Mail Order

♥ Kleer
(see Green Mountain)

♥ Kolestral
(see Wella)

▼ Klorin
(see Colgate-Palmolive)

▼ Kolynos
(see Colgate-Palmolive)

♡ Kmart Corporation ●
3100 W. Big Beaver Road
Troy, MI 48084
(810)643-1000
Personal care, Household

▼ Kotex
(see Kimberly-Clark)

♥ Kryolan Corporation ■ ●
132 9th Street
San Francisco, CA
94103-2603
Cosmetics, Personal care

♥ KMS Haircare Products
(see KMS Research)

♥ KMS Research, Inc. ■
4712 Mountain Lakes Blvd.
Redding, CA 96003
(800)DIALKMS,
(916)244-6000
Personal care

♥ KSA Jojoba ☆ ■
19025 Parthenia Street
Northridge, CA 91324
(818)701-1534
*Cosmetics, Personal care,
Companion animal*

♥ Kneipp Corporation of
America ☆ ■ ✉
Valmont Industrial Park
675 Jaycee Drive
West Hazleton, PA 18201
(800)937-4372
Personal care

♥ L'anza Research
International ☆ ■
935 West Eighth Street
Azusa, CA 90172
(800)423-0307,
(818)334-9333
Personal care

♥ Knowing
(see Estee Lauder)

▼ L'effluer
(see Coty)

💜 L'Herbier De Provence,
Ltd. ☆ ●
462 Fashion Avenue
Floor 17
New York, NY 10018-7606
(212)967-5980
Cosmetics, Personal care

💗 L'Heure Bleue
(see Guerlain, Inc.)

💗 L'Oreal of Paris ■
575 5th Avenue
New York, NY 10017
(212)818-1500
Cosmetics, Personal care

💜 L-Form Amino Acids for
Pets/People
(see Pro-Tec Pet Health)

💜 La Costa Products
International ✉
2875 Loker Ave., East
Carlsbad, CA 92008
(800)-LA-COSTA,
(619)438-2181
Cosmetics, Personal care

💜 La Costa Spa
(see La Costa Products Int'l.)

▼ La Coupe
(see Platex Care)

💜 La Crista, Inc. ☆ ■
P.O. Box 240
Davidsonville, MD 21035
(800)888-2231,
(410)956-4447
Cosmetics, Personal care

▼ La Croix
(see Colgate-Palmolive)

💜 La Dove, Inc. ■
16100 NW 48th Avenue
Hialeah, FL 33014
(305)624-2456
Cosmetics, Personal care

💜 LA Looks Hair Care
(see Dep Corporation)

💜 La Natura ■
425 N. Bedford Drive
Beverly Hills, CA 90210
Personal care

💜 La Parfumerie, Inc. ■
750 Lexington Avenue,
Floor 16
New York, NY 10022-1200
(212)750-1111
Cosmetics, Personal care

💜 Lady Ashford Bath &
Body Products
(see Bio-Tec Cosmetics)

☆ Products contain no
animal or animal
derived ingredients   ■ Manufacturer   ● Distributor   ✉ Mail Order

💜 Lady Burd Exclusive
Private Label Cosmetics ■
44 Executive Blvd.
Farmingdale, NY 11735
(516)454-0444
Cosmetics, Personal care

💜 Lady In Red, Ltd. ☆ ■
P.O. Box 100
East Norwich, NY 11732
(800)LADYRED,
(516)624-8100
Cosmetics, Personal care

💜 Lady of the Lake
Company ☆ ■ ✉
P.O. Box 7140
Brookings, OR 97415
(503)469-3354
Personal care

▼ Lagerfield
(see Unilever U.S.)

💜 Laguna Soaps
(see Clearly Natural Products)

💜 Lakon Herbals ☆ ■ ✉
RR 1, Box 4710
Montpelier, VT 05602
(802)223-5563
Personal care

💜 Lan-O-Sheen, Inc. ☆ ■
289 5th Street, East
Suite 210
St. Paul, MN 55101
(612)293-0522
Household, Personal care

💜 Lancaster, Inc.
(see Revlon)

♡ Lancome
(see L'Oreal of Paris)

💜 Lanex
(see Carma Labs)

💜 Lange Laboratories ■
21093 Forbes Avenue
Hayward, CA 94545
(510)785-6570
Personal care, Cosmetics

▼ Lasting Color by
Loving Care
(see Clairol)

▼ Lasting Color Lipstick
(see Max Factor)

💜 Lasting Pride Cat Litter
(see Oil-Dri)

❤ Laticare
(see Stiefel Labs)

❤ Lauder for Men
(see Estee Lauder)

❤ Laundro-Kleen
(see Wachters')

▼ Laura Biagiotti-Roma
(see Procter & Gamble)

♡ Lauren by Ralph Lauren
(see L'Oreal of Paris)

▼ Lava
(see Procter & Gamble)

❤ Lavilin Deodorant
(see Micro Balanced
Products)

❤ Lavoris Oral Rinse
(see Dep Corporation)

❤ Lazer Pre-Seal
(see Alpha 9)

❤ Le Crystal Naturel
(see French Transit, Ltd.)

▼ Le Jardin
(see Max Factor)

❤ Le Stick
(see Nature de France)

❤ Leaf, Inc.
(see Finnfoods)

▼ Lectric Shave
(see SmithKline Beecham)

♡ Leichner's Stage Make-
Up/Costume House ●
284 King S.W.
Toronto, Ontario
Canada M5V 1J2
Cosmetics

❤ Les Femmes, Inc. ☆ ■ ●
17890 Isle Ave, Ste #100
Lakeville, MN 55044
(612)892-7990
Personal care

♡ Les Meteorites
(see Guerlain, Inc.)

▼ Lestoil
(see Procter & Gamble)

▼ Lever 2000
(see Lever Brothers)

▼ Lever Brothers Company
390 Park Avenue
New York, NY 10022
(800)451-6679

💜 Levlad Inc. ■
9200 Mason Avenue
Chatsworth, CA 91311
(800)327-2012,
(818)882-2951
Cosmetics, Personal care

▼ Lewis Red Devil Lye
(see Reckitt & Colman)

💜 Leydet Aromatics ☆ ■ ● ✉
P.O. Box 2354
Fair Oaks, CA 95628
Cosmetics, Personal care

💜 Li'l Saurys Sunblock
(see Finley Companies)

💜 Life Expander
(see General Nutrition)

💜 Life Tree Products
(see Sierra Dawn Products)

▼ Lifebuoy
(see Lever Brothers)

💜 Lifeline Company ☆ ■
P.O. Box 531
Fairfax, CA 94930
(415)457-9024
Household

💜 Lifestyle
(see La Dove)

💜 Lift-Off
(see Delby System)

▼ Light & Natural Whipped
Creme Make-up
(see Max Factor)

💜 Light Mountain Soap
(see Siddha Int'l)

💜 Light Touch
(see Lotus Light Ent.)

▼ Lightdays
(see Kimberly-Clark)

💜 Liken
(see Earth Science)

💜 Lilt
(see Dep Corporation)

💙 Lily of Colorado ☆ ■ ✉
P.O. Box 12471
Denver, CO 80212
(303)455-4194
Cosmetics, Personal care

♡ Lime Drops Bath Oil
(see House of Lowell)

♡ Lime-O-Sol ■
State Rd., 4, P.O. Box 395
Ashley, IN 46705
(219)587-9151
Household

💙 Limited Stores
(see Gryphon Development)

💙 Line Tamer
(see Earth Science)

💙 Lip Apeel
(see BeautiControl Cosmetics)

💙 Lip Savvy
(see Strong Skin Savvy)

💙 Lip Smackers
(see Bonne Bell)

♡ Lip Trip
(see Mountain Ocean)

💙 Lipservice Facial
Hair Removal
(see Les Femmes)

♡ Liquid Copper Glo
(see SerVaas Labs)

▼ Liquid Glory
(see S.C. Johnson & Son)

▼ Liquid Paper
(see Gillette)

▼ Liquid Plumr
(see Clorox)

▼ Litter Green
(see Clorox)

💙 Little Herbal Garden ☆ ■ ✉
33 Elm Street
Roslyn Heights, NY 11577
(516)625-0443
Cosmetics, Personal care

💙 Living Earth Bath Treatment
(see Norimoor)

♡ Liz Claiborne Cosmetics ■ ●
1441 Broadway
New York, NY 10018
(212)354-4900
Cosmetics, Pesonal care

♥ Loanda Herbal Soaps
(see Carme)

♥ Lobob Laboratories, Inc. ■
1440 Atteberry Lane
San Jose, CA 95131-1410
(800)835-6262,
(408)432-0580
Personal care

♥ Logics
(see Matrix Essentials)

♥ Logona USA, Inc. ■
554-E Riverside Drive
Asheville, NC 28801
(704)252-1420
Cosmetics, Personal care

♥ Lotions & Potions ■ ● ✉
422 South Mill Avenue
Tempe, AZ 85281
(800)462-7595,
(602)968-4652
Cosmetics, Personal care

♥ Lotus Brands ☆ ●
P.O. Box 325
Twin Lakes, WI 53181
(414)889-8561
*Cosmetics, Personal care,
Household*

♥ Lotus Light Enterprises ☆ ●
P.O. Box 1008
Silver Lake, WI 53170
(800)548-3824, (414)889-8501
Cosmetics, Personal care

♥ Lotus Pads ☆ ■
131 NW Fourth, Suite 156
Corvallis, OR 97330
(503)758-4110
Personal care

♥ Louise Bianco Skin
Care, Inc. ✉
13655 Chandler Blvd.
Sherman Oaks, CA 91401
(800)782-3067,
(818)786-2700
Personal care

♡ Love Frangrances
(see Mem Company)

♥ Love Mitts
(see Body Love Natural
Cosmetics)

♡ Love's Products
(see Mem Company)

▼ Loving Care
(see Clairol)

♡ LT York Company ☆ ■
P.O. Box 50
Bucklin, MO 64631
Personal care

❤ Lucille Flint Formulas
(see Kallima Int'l)

❤ Lucky Kentucky
(See Palm Beach
Beauty Products)

♡ Lucky Tiger
(see L.T. York)

❤ Lume International ■
1692 S. Chambers Road
Aurora, CO 80017-5058
Personal care

❤ Lunar Farms Herbal
Specialist ■ ●
3 Highland-Greenhills
Gilmer, TX 73644
(903)734-5893
Personal care

▼ Lustrasilk
(see Gillette)

▼ Luvs
(see Procter & Gamble)

▼ Lux
(see Unilever U.S.)

❤ Luzier Personalized
Cosmetics ■ ✉
3216 Gillham Plaza
Kansas City, MO 64109
(816)531-8338
Cosmetics, Personal care

❤ Lynda Charles ✉
Cosmetics, Int'l.
3360 Wiley Post, Ste. C
Carrollton, TX 75006
(800)228-3215
Cosmetics, Personal care

❤ M Cologne/After Shave
(see Marilyn Miglin L.P.)

❤ MAC Cosmetics
(see Make-Up Art
Cosmetics Ltd.)

❤ Macleans
(see SmithKline Beecham)

❤ Magic Cream
(see Lunar Farms
Herbal Specialists)

▼ Magic Mushroom
(see Reckitt & Colman)

☆ Products contain no   ■ Manufacturer   ● Distributor   ✉ Mail Order
animal or animal
derived ingredients

💜 Magic of Aloe, Inc. ■ ● ✉
7300 N. Crescent Blvd.
Pennsauken, NJ 08110
(800)257-7770, (609)662-3334
Cosmetics, Personal care

💜 Magic Sizing
(see Dial Corporation)

💜 Magical Mane
(see Conair)

▼ Magical Musk
(see Minto Industries)

💜 Magick Botanicals/
Magick Mud ☆ ■
3412 West MacArthur Blvd.
Suites J-K
Santa Ana, CA 92704
(800)237-0674,
(714)957-0674
Personal care

💜 Magiclean
(see Minto Industries)

💜 Maharishi Ayur-Veda
Products ● ✉
P.O. Box 49667
Colorado Springs, CO
80949-9667
(719)260-5500
Cosmetics, Personal care

💜 Makamina Inc. ☆ ● ✉
P.O. Box 307
Wallingford, PA 19086
(800)537-6746
Personal care

💜 Make-Up Art
Cosmetics Ltd. ☆ ■ ●
233 Carlton Street, #201
Toronto, Ontario
Canada M5A 2L2
(800)387-6707,
(416)924-0598
Cosmetics, Personal care

💜 Mallory Pet Supplies ☆ ■
126 Atrisco Place, SW
Alburquerque, NM 87105
(505)836-4033
Companion animal

💜 Mane Street Hair
Products ☆ ■
7201 York Avenue, S.
Edina, MN 55435
(612)832-0061
Personal care

💜 Manpower
(see Dial Corporation)

💜 CRUELTY FREE
Does not test products
or ingredients on animals

💜 Ingredients MAY
be tested on
animals

▼ Tests products or
ingredients
on animals

❤ Mar-Riche Enterprises,
Inc. ☆ ■ ●
640 Glass Lane
Modesto, CA 95356
(209)529-1757
Cosmetics, Personal care

❤ Marcal Paper Mills, Inc. ☆ ■
1 Market Street
Elmwood Park, NJ 07407
(201)796-4000
Household

❤ Marcha Labs, Inc. ■
P.O. Box 186
Terry, MT 59349
Personal care

▼ Marche Image Corporation
P.O. Box 1010
Bronxville, NY 10708
(800)753-9980,
(914)793-2093
Cosmetics, Personal care

❤ Margarite Cosmetics/
Moon Products, Inc. ■
2138 Okeechobee Blvd.
W. Palm Beach, FL 33409
Personal care

❤ Marilyn Miglin Cosmetics
(see Marilyn Miglin L.P.)

❤ Marilyn Miglin,
L.P. ☆ ■ ● ✉
112 East Oak Street
Chicago, IL 60611
(800)662-1120, (312)943-1120
Cosmetics, Personal care

❤ Mario Badescu Skincare,
Inc. ☆ ■ ✉
320 E. 52nd Street
New York, NY 10022
(212)758-1065
Personal care, Cosmetics

❤ Martin Von Myering
Inc. ■ ● ✉
422 Jay Street
Pittsburgh, PA 15212
(412)323-2832
Personal care

❤ Mary Kay Cosmetics, Inc. ■
8787 Stemmons Freeway
Dallas, TX 75247
(800)MARYKAY
*Cosmetics, Personal care,
"Moratorium"*

❤ Masada Marketing
Company ☆ ■
P.O. Box 4767
N. Hollywood,CA 91607
(800)368-8811, (818)503-4611
Personal care

☆ Products contain no   ■ Manufacturer   ● Distributor   ✉ Mail Order
animal or animal
derived ingredients

♥ Massage Essence
(see Wachters')

▼ Massengill
(see SmithKline Beecham)

♥ Master's Flower
Essences ☆ ■ ● ✉
14618 Tyler Foote Road
Nevada City, CA 96969
Personal care

♥ Mastey De Paris, Inc. ■
25413 Rye Canyon Road
Valencia, CA 91355
(800)6MASTEY,
(805)257-4814
Cosmetics, Personal care

▼ Matchabelli
(see Prince Matchabelli)

♥ Mate
(see Melaleuca)

♥ Matrix Essentials, Inc. ☆ ■
30601 Carter Street
Solon, OH 44139
(800)282-2822,
(216)248-3700
Cosmetics, Personal care

▼ Max Factor & Co.
(see Procter & Gamble)

▼ Maxi
(see Max Factor)

▼ McGregor
(see Chesebrough-Pond's)

♥ McLaughlin, Inc. ☆ ■ ● ✉
847 N. Second Avenue
Suite 120
New York, NY 10017
(212)682-6437
Household, Personal care

♡ Mechanics Brand
(see Blue Coral)

♥ Medi-Feet
(see American Cosmetics Ind.)

♡ Medi-Kay
(see L.T. York)

♥ Medical
(see Palm Beach Beauty
Products)

♥ Medicine Flower ☆ ■ ● ✉
720 NE Granger Avenue
Suite A
Corvallis, OR 97330-9660
(503)745-3055
Cosmetics, Personal care

💜 MediPatch Laboratories
Corporation ☆ ■
Box 795
E. Dennis, MA 02641
(508)385-4549
Companion animal

💜 Mediterranean
(see Delby System)

💜 Melaleuca, Inc. ■
3910 South Yellowstone
Idaho Falls, ID 83402
(208)522-0700
*Cosmetics, Personal care,
Household, Companion
animal*

▼ Meltonian
(see Kiwi Brands)

♡ Mem Company, Inc. ■
Union Street
Northvale, NJ 07647-0928
(201)767-0100
Cosmetics, Personal care

💜 MEN by Geoff
Thompson ■ ✉
6663 SW Beaverton
Hillsdale Hwy, #281
Portland, OR 97225-1403
(503)257-7066
Personal care

▼ Men's Choice
(see Clairol)

▼ Men's Fragrance
(see Chesebrough-Pond's)

♡ Menley & James Labs,
Inc. ☆ ●
100 Tournament Drive
Horsham, PA 19044
(215)441-6500
Cosmetics, Personal care

▼ Mennen Company
East Hanover Avenue
Morristown, NJ 07962-1928
(201)631-9000

💜 Mentholatum Company
of Canada, Ltd. ■
16 Lewis Street
Fort Erie, Ontario
Canada L2A 5M6
(416)871-1665
Personal care

💜 Mer-Flu-An
(see Compassion Matters)

💜 Mera Personal Care
Products ☆ ■ ✉
P.O. Box 218
Circle Pines, MN 55014
(800)752-7261
Personal care

💜 Merce Gelle'
(see Key Distributors)

💜 Mere Cie Inc. ☆ ■ ●
1100 Soscol Rd., #3
Napa, CA 94558
(800)832-4544,
(707)257-8510
Personal care, Household

💜 Merle Norman Cosmetics ■
9130 Bellanca Avenue
Los Angeles, CA 90045
(800)421-2060,
(213)641-3000
Cosmetics, Personal care

▼ Mersena Denture Cleanser
(see Colgate-Palmolive)

▼ Mersene
(see Colgate-Palmolive)

💜 Meta International
(see Dena Corp.)

💜 Metrin Laboratories
Ltd. ☆ ■ ● ⊠
4360 Rockridge Road
West Vancouver, BC
Canada V7W 1A7
(604)922-8111
*Cosmetics, Personal care,
Companion animal*

▼ MFC
(see Mennen)

▼ MFR Cosmetics
(see Elizabeth Arden)

💜 Mia Rose Products, Inc. ☆ ■
177-F Riverside Ave.
Newport Beach, CA 92663
(800)292-6339,
(714)662-5465
Household

💜 Michael's Naturopathic
Programs ■
6820 Alamodowns Pkwy.
San Antonio, TX 78238
(800)525-9643,
(210)647-4700
Personal care

💜 Michel Constantini
Cosme ● ⊠
124 West 72nd Street
New York, NY 10023
(212)501-8245
Cosmetics, Personal care

💜 Micro Balanced
Products ☆ ■ ⊠
25 Aladdin Ave.
Dumont, NJ 07628
(800)626-7888, (201)387-0200
Personal care

💜 CRUELTY FREE
Does not test products
or ingredients on animals

💜 Ingredients MAY
be tested on
animals

▼ Tests products or
ingredients
on animals

❤ Microderm
(see BeautiControl
Cosmetics)

❤ Mild & Natural Childrens
Products
(see Carme)

▼ Miles Inc.
P.O. Box 340
Elkhart, IN 46515
(219)264-8111

❤ Mill Creek
(see Carme)

▼ Mink Difference
(see Gillette)

❤ Mint Julep
(see Para Labs)

❤ Minto Industries Ltd. ☆ ■
54 Granville Road
London SW18 5SQ
Great Britain
*Personal care, Household,
Companion animal*

▼ Miracle White
(see Kiwi Brands)

▼ Miss Clairol
(see Clairol)

❤ Miss Dior
(see Chrsistian Dior
Perfumes)

❤ Mission Hills ■
21093 Forbes Avenue
Hayward, CA 94545
(510)785-6570
Personal care

♡ Mitsouko
(see Guerlain, Inc.)

❤ MLE Essential Essence Oils
(see Wachters')

❤ Moby Oil
(see J & J Jojoba)

❤ Modafini Inc. ☆ ■
P.O. Box 6870
Malibu, CA 90265
(800)800-6632,
(805)520-7050
Personal care

❤ Moderations
(see Matrix Essentials)

❤ Modern Research
(see Image Labs)

❤ Moisture Maximizer
(see Alpha 9)

☆ Products contain no   ■ Manufacturer   ● Distributor   ⊠ Mail Order
animal or animal
derived ingredients

▼ Moisture Rich Lipstick
New Definition Lip Color
(see Max Factor)

♥ Moisture Therapy
(see Avon Products)

♥ MoisturEyes
(see Carme)

♥ Moisturing Hand Creme
(see Wachters')

♥ Molinard Fragrance and
Bath Lines
(see Dep Corporation)

♥ Monoi
(see Hawaiian Resources)

♥ Montagne Jeunesse ■ ●
The Old Grain Store
4 Denne Rd.
Horsham, West Sussex
England RH12 1JE
01-40-327-2737
Personal care

♥ Moondrops
(see Revlon)

♥ Moonshadow Carwash
Concentrate
(see Earthly Matters)

♥ Moriah
(see Colora)

♥ Morrill's New Directions ✉
P.O. Box 30
Orient, ME 04471
(800)368-5057
Companion animal

▼ Most Precious
(see Unilever United States)

♥ Most Products Inc. ■ ●
3604 Edinburgh Drive
Kalamazoo, MI 49007
(800)331-6678, (616)381-6678
Cosmetics, Personal care

♥ Mother Love Herbal Co. ■
280 Stratton Park
Bellvue, CO 80512
(303)493-2892
Personal care

♥ Mother's Fragrances,
Incense & Perfume Oils
(see Mere Cie)

♥ Mother's Little Miracle,
Inc. ☆ ●
930 Indian Peak Rd., Ste. 215
Rolling Hills Estates, CA 90274
(310)544-7125
Personal care, Household

♥ CRUELTY FREE
Does not test products
or ingredients on animals

♥ Ingredients MAY
be tested on
animals

▼ Tests products or
ingredients
on animals

♡ Mother's Special Blend
(see Mountain Ocean)

♥ Mothers Little Helper All
Purpose Cleaner
(see Earthly Matters)

▼ Motif
(see Clairol)

♥ Mountain Fresh
(see Dial Corporation)

♥ Mountain Herbery
(see Carme)

♡ Mountain Ocean Ltd. ■
Box 951
Boulder, CO 80306
(303)444-2781
Personal care

♥ Mountain Rose Herbs ■ ✉
P.O. Box 2000
Redway, CA 95560
(707)923-3941
*Cosmetics, Personal care,
Companion animal*

♥ Mountain Streams Acid
Free Drain Opener
(see Earthly Matters)

♥ Movie Stars
(see Delby Systems)

♥ Mr. Christal's Inc. ☆ ■
1100 Glendon Ave, Ste 1250
Los Angeles, CA 90024
(800)426-0108,
(310)824-2508
Companion animal

♥ Mr. Christal's Luxury
Australian Shampoo
(see Mr. Christal's)

▼ Mr. Clean
(see Procter & Gamble)

▼ Mr. Muscle
(see S.C. Johnson & Son)

♥ Muelhens Inc. ☆ ■
15 Executive Blvd.
Orange, CT 06477
(800)243-5555
(203)787-4711
Cosmetic, Personal care

♥ Murad, Inc. ☆ ■
2121 Rosecrans Avenue
5th Floor
El Segundo, CA 90245
(800)242-1103
Personal care

☆ Products contain no
animal or animal
derived ingredients   ■ Manufacturer   ● Distributor   ✉ Mail Order

▼ Murphy's Oil Soap
(see Murphy-Phoenix Co.)

▼ Murphy-Phoenix Company
25800 Science Park Drive
Beachwood, OH 44122
(800)486-7627

♡ Musk
(see Mem Corporation)

♥ Nadina's Cremes ☆ ■
3600 Clipper Mill, Suite 140
Baltimore, MD 21211
(800)722-4292,
(410)235-9192
Personal care

♥ Nail Dry
(see Gena Labs)

▼ Nail Thick
(see Max Factor)

♥ Nailogics
(see BeautiControl
Cosmetics)

▼ Namel Dry
(see DowBrands, Inc.)

♡ Narwhale Of High
Tor Ltd. ■ ✉
591 South Mountain Rd.
New City, NY 10956
(800)354-2407,
(914)358-5743
Cosmetics, Personal care

♡ National Home Care
Products, Inc. ■
2075 West Scranton Ave.
Porterville, CA 93257
(209)781-8871
Cosmetics, Personal care

♥ Native Scents, Inc. ☆ ■
Box 5639
Toas, NM 87571
(505)758-9656
Personal care

♥ Natracare ☆ ■
191 University Blvd.
Suite 294
Denver, CO 80206
(800)796-2872,
(303)320-1510
Personal care

♥ Natura
(see Goodebodies)

❤ Naturade ■ ●
7110 East Jackson Street
Paramount, CA 90723-4895
(310)531-8120
Personal care

❤ Naturade Aloe Vera 80
Collection
(see Naturade)

❤ Natural Alternative
(see Pet Connection)

❤ Natural Animal, Inc. ☆ ■
7000 US 1 North
P.O. Box 1177
St. Augustine, FL 32095
(800)274-7387,
(904)824-5884
Companion animal

❤ Natural Apricot Facial Wash
(see Wachters')

❤ Natural Attitudes, Inc. ☆ ●
10320 Camino Santa Fe
Suite A
San Diego, CA 92121
(619)623-1340
Personal care

❤ Natural Baby Soap
(see Wachters')

❤ Natural Bodycare ☆ ■ ⊠
P.O. Box 249
Camarillo, CA 93012
(805)482-7791
Personal care, Household

❤ Natural Care
(see Conair)

❤ Natural Chemistry ☆ ■
244 Elm Street
New Canaan, CT 06840
(800)753-1233,
(203)966-8761
*Household, Companion
animal*

❤ Natural Color Cosmetic
(see Aveda Corp.)

❤ Natural Enzyme
Formula for Baquacil
(see Natural Chemistry)

❤ Natural Essence Soaps
(see Clearly Natural
Products)

♡ Natural Glow ■ ●
565 Broad Hollow Road
Farmingdale, NY 11735
(516)293-7070
Cosmetics, Personal care

☆ Products contain no   ■ Manufacturer   ● Distributor   ⊠ Mail Order
animal or animal
derived ingredients

▼ Natural Instincts
(see Clairol)

💜 Natural Nectar
Corporation ☆ ■
16010 Phoenix Drive
Hacienda Heights, CA
91745-1623
Personal care

💜 Natural Research
People Inc. ☆ ■
South Route, Box 12
Lavina, MT 59046
(406)575-4343
Companion animal

💜 Natural Soap
(see Dial Corporation)

💜 Natural Therapeutics
Centre ☆ ■ ●
2500 Side Cove
Austin, TX 78704
(512)444-2862
Personal care

💜 Natural Way Natural
Body Care ■ ● ✉
820 Massachusetts Street
Lawrence, KS 66044
Personal care

💜 Natural Wonder
(see Revlon)

💜 Natural World, Inc. ✉ ●
7373 N. Scottsdale Road
Suite A-280
Scottsdale, AZ 85253-3550
(800)728-3388
*Cosmetics, Personal care,
Household, Companion
animal*

💜 Naturally Free, The
Herbal Alternative
(formerly Skedaddle,
see Bug-Off, Inc.)

💜 Naturally Yours, Alex ☆ ■ ✉
P.O. Box 3398
Holiday, FL 34698
(813)733-9592
Companion animal

💜 Nature 1st
(see Natural Research
People)

💜 Nature Clean
(see Frank T. Ross &
Sons, Ltd.)

💜 Nature Cosmetics
(see Avanza)

💜 CRUELTY FREE
Does not test products
or ingredients on animals

💜 Ingredients MAY
be tested on
animals

▼ Tests products or
ingredients
on animals

❤ Nature de France, Ltd.
(see Para Labs)

❤ Nature Food Centres, Inc. ☆ ■
921 Penn Avenue
Pittsburgh, PA 15222-3814
(508)657-5000
Cosmetics, Personal care

❤ Nature Fresh
(see Thursday Plantation Labs)

❤ Nature's Acres ☆ ■ ● ⊠
East 8984 Weinke Rd.
North Freedom, WI 53951
(608)522-4492
Cosmetics, Personal care

❤ Nature's Best
(see Natural Research
People, Inc.)

❤ Nature's Colors
Cosmetics ☆ ■ ●
424 La Verne Avenue
Mill Valley, CA 94941
(415)388-6101
Cosmetics, Personal care

❤ Nature's Corner ■ ● ⊠
P.O. Box 4111
Alvin, TX 77512
(713)388-0740
Personal care

❤ Nature's Country Pet ☆ ● ⊠
1765 Garnet Avenue
Suite 12
San Diego, CA 92109
(800)576-PAWS
Companion animal

❤ Nature's Crystal
(see Deodorant Stones of
America)

❤ Nature's Elements
International, Ltd. ☆ ■ ●
115 River Road
Edgewater, NJ 07020-1009
(201)945-2640
Cosmetics, Personal care

❤ Nature's Family Skin Care
(see Dep Corporation)

❤ Nature's Gate
(see Levlad)

❤ Nature's Key
(see American Eco-Systems)

❤ Nature's Manicure
(see Gena Labs)

❤ Nature's Miracle
(see Pets 'N People)

☆ Products contain no
animal or animal
derived ingredients
■ Manufacturer ● Distributor ⊠ Mail Order

💜 Nature's Pearl
(see Deodorant Stones
of America)

💜 Nature's Pharmacy ☆ ■
P.O. Box 552
Soquel, CA 95073
(408)438-1700
Personal care

💜 Nature's Plus ■
548 Broadhollow Road
Melville, NY 11747
(800)645-9500,
(516)293-0030
Personal care, Companion animal

💜 Nature's Plus, The
Energy Supplements
(see Nature's Plus)

💜 Nature's Pure Crystal
Deodorant
(see Hargen Distributors)

💜 Nature's Rescue
(see Ellon USA)

💜 Nature's Sunshine
Products, Inc. ☆ ■
P.O. Box 19005,
75 East 1700 South
Provo, UT 84605-9005
(801)342-4300
Personal care, Cosmetics

💜 NaturElle Cosmetics
Corporation ■ ● ✉
P.O. Box 9
Pine, CO 80470-0009
(800)442-3936
Cosmetics, Personal care

💜 Naturessence
(see Avanza)

💜 NatureWorks Inc. ■
5341 Dairy Avenue
Suite F & G
Agoura Hills, CA 91301
(800)843-9535, (818)889-1602
Personal care

♡ Naturistics ■ ●
565 Broad Hollow Road
Farmingdale, NY 11735
(516)293-7070
Personal care

💜 Naturotherapy Bath &
Body Products
(see Essential Aromatics)

❤ Natus ✉
4550 W.77th St. #300
Edina, MN 55435
(800)833-4627,
(612)835-2317
Personal care

▼ Navy
(see Procter & Gamble)

❤ Nectar USA ☆ ■ ●
4164-A Inns Lake
Glen Ellen, PA 23060
(804)527-4356
Cosmetics, Personal care

❤ Nectarine
(see AKA Saunders, Inc.)

❤ Neil's Yard
(see LaNatura)

❤ Nemesis Inc. ☆ ■
4525 Hiawatha Ave. South
Minneapolis, MN 55406
(612)724-5732
Personal care

♡ Nenna Gold
(see Shikai Products)

❤ New Age Creations/
Jeanne Rose ■ ✉
219 Carl Street
San Francisco, CA 94117
(415)564-6785
*Personal care, Companion
animal*

❤ New Beginnings
(see Natural Bodycare)

❤ New Cycle Products,
Inc./Womankind ☆ ■ ✉
P.O. Box 1775
Sebastopol, CA 94973
(707)829-2744
Personal care

▼ New Definition
Perfecting Make-Up
(see Max Factor)

❤ New Direction ☆ ■ ●
300 Country Club Road
Avon, CT 06001
(203)675-9401
*Personal care, Companion
animal*

❤ New Direction
Shampoo Bar
(see New Direction)

▼ New Freedom
(see Kimberly-Clark)

❤ New Methods ☆ ● ✉
P.O. Box 22605
San Francisco, CA 94122
(415)664-3469
Companion animal,
Household

❤ New Moon Extracts,
Inc. ☆ ■ ●
99 Main Street
Brattleboro, VT 05301
(800)543-7279,
(802)257-0018
Cosmetics, Personal care

♡ New Zealand Hair Paradise
(see Redmond Products)

❤ Neways, Inc. ☆ ■ ● ✉
150 E. 400 N.
Salem, UT 84653
(800)799-5656,
(801)423-2800
Cosmetics, Personal care

❤ Nexxus Products Company ●
P.O. Box 1274
Santa Barbara, CA 93116
(800)444-6399,
(805)968-6900
Personal care

❤ Nice 'N Thick
(see Palm Beach Beauty
Products)

▼ Nice'n' Easy
(see Clairol)

❤ Nino Cerutti Fragrance
and Grooming for Men
(see Dep Corporation)

❤ Nirvana Inc. ☆ ■
P.O. Box 18413
Minneapolis, MN 55418
(612)932-2919
Personal care

❤ Nite Caress Night Creme
(see Wachters')

❤ Nivea
(see Beiersdorf, Inc.)

▼ No Color Mascara
(see Max Factor)

❤ No Sweat
(see Revlon)

▼ Non-Clay
428 South Forest
Denver, CO 80222
(303)377-4000

❤ Norelco Consumer
Products Company ■
P.O. Box 10166
Stamford, CT 06904
(203)329-2400
Personal care

❤ Norell Perfumes Inc.
(see Revlon)

❤ Norimoor Company,
Inc. ☆ ■ ●
La Maison Francaise
Fifth Ave., #5222
New York, NY 10185
Personal care

❤ North Country Glycerin
Soap and Candle ■
7888 County Road #6
Maple Plain, MN 55359-9552
(800)328-4827 ext. 2153,
(612)479-3381
*Personal care, Companion
animal*

❤ Nourishair
(see General Nutrition)

▼ Noxell Corporation
(see Procter & Gamble)

▼ Noxema
(see Procter & Gamble)

▼ Noxon Metal Polish
(see Reckitt & Colman)

▼ Nucleic A
(see DowBrands)

❤ Nucleic Plus
(see Palm Beach Beauty
Products)

❤ Nutra Cleanse
(see Espial)

❤ Nutri-Cell, Inc. ☆ ■
1038 N. Tustin, Ste. 309
Orange, CA 92667-5958
(714)953-8301
Personal care

❤ Nutri-Metics Intl., Inc. ☆ ■
12723 East 166th Street
Cerritos, CA 90701
(310)802-0411
Cosmetics, Personal care

❤ Nutribiotic ■
865 Parallel Drive
Lakeport,CA 95453
(707)263-0411
Personal care

▼ o.b. Tampons
(see Johnson & Johnson)

☆ Products contain no   ■ Manufacturer ● Distributor ✉ Mail Order
  animal or animal
  derived ingredients

♥ Oasis Biocompatible ☆ ■
5 San Marcos Trout Club
Santa Barbara, CA 93105
(805)967-3222
Household

♥ Oasis Brand Products ☆ ■
P.O. Box 12871
La Jolla, CA 92039
(619)276-1440
Personal care

▼ Octagon Dishwashing
Liquid
(see Colgate-Palmolive)

♡ Odelys
(see Guerlain, Inc.)

♥ Odor-B-Gone
(see Gannon's)

▼ Off!
(see S.C. Johnson & Son)

♡ Oil Free Hand & Body
Lotion
(see Botanicus)

▼ Oil of Olay
(see Procter & Gamble)

♥ Oil-Dri Corp. of
America ☆ ■ ●
410 N. Michigan Ave.
Chicago, IL 60611
(312)321-1515
Companion animal

♥ OJA-Skin, Hair &
Body Care
(see The Little Herbal
Garden)

▼ Old Spice
(see Procter & Gamble)

♥ Olde Tyme 1881
(see Green Mountain)

♥ Oleg Cassini
(see Cassini Parfums)

♥ Oliv
(see Shahin Soap)

♡ Oliva
(see Home Health Products)

♥ Oliva Stoard
(see Natural Bodycare)

♥ One 'N Only
(see Conair)

▼ One Step
(see S.C. Johnson & Son)

▼ One Touch
(see Johnson & Johnson)

♥ One Unlimited Perfume
(see La Parfumerie)

♥ One World Botanicals,
Ltd. ☆ ■ ● ✉
326 Broad Street
Red Bank, NJ 07701
(908)530-2011
Companion animal

♥ OP-Tics
(see I-Tech Labs.)

♥ OPI Products, Inc. ☆ ■
13056 Saticoy Street
N. Hollywood, CA 91605
(800)341-9999,
(818)759-2400
Cosmetics, Personal care

♥ Opticurl
(see Matrix Essentaisl)

▼ Option
(see Clairol)

▼ Orafix Denture Adhesive
(see SmithKline Beecham)

▼ Oral B
(see Gillette)

♡ Oral Logic, Inc. ■
7000 Burdick Exp. E
Minot, ND 58701
(800)345-1143
Personal care

▼ Orancin
(see Procter & Gamble)

♥ Orange Plus
(see Earth Friendly
Products)

♥ Orange-Magic
(see Blue Ribbon Pet Care)

♥ Orange-Mate Inc. ☆ ■
P.O. Box 883
Waldport, OR 97394
(800)626-8685,
(503)563-3290
Personal care, Household

♥ Orchard Mist
(see Melaleuca)

♥ Organic Moods
(see KMS Research)

♥ Organimals
(see Aubrey Organics)

☆ Products contain no   ■ Manufacturer   ● Distributor   ✉ Mail Order
animal or animal
derived ingredients

💜 Original Ojai Flower Candle
(see Essential Aromatics)

💜 Origins Natural
Resources Inc.
(see Estee Lauder)

♡ Orjene for Men
(see Orjene Natural
Cosmetics)

♡ Orjene Natural
Cosmetics Co., Inc. ■
543 48th Avenue
Long Island City, NY
11101-5694
(718)937-2666
Personal care

💜 Orlane Inc. ■ ●
555 Madison Avenue
New York, NY 10022
(212)750-1111
Cosmetics, Personal care

♡ Orly International ☆ ■
9309 Deering Avenue
Chatsworth, CA 91311
(818)998-1111
Personal care

▼ Overnight Moisture
Supplement
(see Max Factor)

▼ Overnight Sucess
(see Max Factor)

▼ Oxy Products
(see SmithKline Beecham)

▼ Oxydol
(see Procter & Gamble)

♡ Oxyfresh USA, Inc. ✉
E. 12928 Indiana Avenue
Spokane, WA 99216-2709
(509)924-4999
Cosmetics, Personal care,
Household, Companion animal

💜 P-Bee Products ☆ ■ ●
8280-123 Clmt. Mesa Blvd.
San Diego, CA 92111
(619)560-7945
Cosmetics, Personal care

💜 Pacific Scents Inc. ☆ ■
P.O. Box 8205
Calabasas, CA 91372
Personal care

💜 Palm Beach Beauty
Products ■
950 Xenia Avenue South
Minneapolis, MN 55416
(800)326-7256, (612)546-0322
Cosmetics, Personal care,
Companion animal

♡ Palma Christi
(see Home Health Products)

▼ Palmolive
(see Colgate-Palmolive)

♡ Paloma Picasso
(see L'Oreal of Paris)

▼ Pam Non-Stick Cooking Spray
(see Reckitt & Colman)

▼ Pampers
(see Procter & Gamble)

▼ Pan Cake & Pan Stick
Make-Up
(see Max Factor)

▼ Pan Handl'rs & Golden
Fleece Cleaning Pads
(see Reckitt & Colman)

▼ Pantene
(see Procter & Gamble)

▼ PaperMate
(see Gillette)

♡ Para Laboratories, Inc. ■
100 Rose Avenue
Hempstead, NY 11550
(800)645-3752, (516)538-4600
Personal care

❤ Paraffin Springs Therapy
Products
(see Gena Labs)

❤ Parfait Cosmetics
(see Blue Cross Beauty
Products)

♡ Parfums Guy Laroche
(see L'Oreal of Paris)

▼ Parfums Int'l., Ltd.
(see Elizabeth Arden)

❤ Parfums Stern
(see Avon)

▼ Parker
(see Gillette)

❤ Parker & Bailey ☆ ■
P.O. Box 12499
Chicago, IL 60612
(312)243-6610
Household

♡ Parlux Fragrances, Inc. ☆ ■
650 SW 16th Terrace
Pompano Beach, FL
33069-4533
(305)946-7700
Personal care

☆ Products contain no   ■ Manufacturer   ● Distributor   ⊠ Mail Order
animal or animal
derived ingredients

💜 Parson's Ammonia
(see Dial Corporation)

💜 Parvenu-Weight Control
(see Espial)

▼ Passion
(see Elizabeth Arden)

💜 Patricia Allison ■ ● ✉
4470 Monahan Road
La Mesa, CA 91941
(800)858-8742, (619)444-4879
Cosmetics, Personal care

💜 Patrick, Ltd.
(see Elysee Scientific
Cosmetics)

💜 Paul Mazzotta, Inc. ☆ ■
P.O. Box 96
Reading, PA 19607
(800)562-1357, (215)376-2250
Cosmetics, Personal care

💜 Paul Mitchell Systems
(see John Paul Mitchell)

♡ Paul Penders Co., Inc. ■
1340 Commerce Street
Petaluma, CA 94954
(800)440-7285, (707)763-5828
Cosmetics, Personal care

💜 Peaceable Kingdom ☆ ✉
1902 West 6th Street
Wilmington, DE 19805
(302)652-7840
*Cosmetics, Personal care,
Household, Companion animal*

💜 Pears Skin Care
(see Dep Corporation)

💜 Pedi Pak
(see Gena Labs)

💜 PediCare
(see American Cosmetics Ind.)

💜 Peelu ☆ ■ ●
109 1/2 Broadway
Fargo, ND 58102
Personal care

💜 Penn Herb Company,
Ltd. ☆ ■ ● ✉
603 N. 2nd Street
Philadelphia, PA. 19123
(215)925-3336
Personal care

💜 Pepper's ☆ ✉
374 Congress Street
Boston, MA 02118
(800)754-6040
Cosmetics, Personal care

▼ Pepsodent
(see Chesebrough-Pond's)

❤ Perfect Balance Skincare
Products
(see Marilyn Miglin L.P.)

❤ Perfect Nail System
(see Gena Labs)

❤ Perfect Natural Products ☆ ■
P.O. Box 13266
Pittsburgh, PA 15243
(412)279-3081
Household, *Personal care*

♡ Performing Preference
(see L'Oreal of Paris)

❤ Perlier Natural Recipes
(see La Parfumerie)

▼ Perma Soft
(see DowBrands)

▼ Pert
(see Procter & Gamble)

❤ Pet Air
(see Mia Rose Products)

❤ Pet Connection ■
P.O. Box 391806
Mountain View, CA 94039
(415)949-1190
Companion animal

❤ Pet Guard, Inc. ☆ ■ ●
165 Industrial Loop S.
Unit #5
Orange Park, IL 32073
(800)874-3221
Companion animal

❤ Pet Guard
(see Basically Natural)

❤ Pet Tech, Inc. ☆ ■
7105 27th Street West
Tacoma, WA 98466
(800)577-9797
Companion animal

❤ Petal Fresh
(see Levlad, Inc.)

❤ Petites
(see Delby System)

❤ Pets 'N' People Inc. ☆ ■
930 Indian Peak Road
Suite 215
Rolling Hills Estate, CA 90274
(310)544-7125
Companion animal

♥ Phantom Frights Cosmetics
(see Blue Cross Beauty
Products)

♥ Pharmagel Corporation ☆ ■
P.O. Box 50531
Santa Barbara, CA 93150
(800)882-4889, (805)568-0022
Cosmetics, Personal care

♥ Pheromone
(see Marilyn Miglin)

♥ Philip B. Inc. ✉
P.O. Box 15341
Beverly Hills, CA 90209
(800)643-5556
Cosmetics, Personal care

♥ Phybiosis ☆ ● ✉
P.O. Box 992
Bowie, MD 20718
(301)805-7920
Cosmetics, Personal care

♡ Pine Glo Products, Inc. ☆ ■
P.O. Box 429
Highway 401
Rolesville, NC 27571
(919)556-7787
Household

▼ Pine Sol
(see Clorox)

♥ Pioneer Brand
(see General Nutrition)

♥ Placenta Plus
(see Palm Beach Beauty
Products)

♥ Plak Attack Dental Rinse
(see Key Distributors)

♥ PlantEssence ■
P.O Box 14743
Portland, OR 97214-0743
(503)621-1312
Household, Personal care

▼ Platinum-Plus Blades
(see Gillette)

▼ Playtex Beauty Care, Inc.
(see Playtex Family
Products Corp.)

▼ Playtex Family Products
Corporation
P.O. Box 728
Paramus, NJ 07653
(800)222-0453

▼ Pledge
(see S.C. Johnson & Son)

♥ Plus 30 Cream
(see Bonne Bell)

♥ CRUELTY FREE
Does not test products
or ingredients on animals

♡ Ingredients MAY
be tested on
animals

▼ Tests products or
ingredients
on animals

💜 Podia Cream
(see Comfort Manufacturing)

♡ Poetry of Flowers
(see Twinscents)

💜 Poison
(see Christian Dior Perfumes)

▼ Polly Berge
(see Chesebrough-Pond's)

♡ Polo by Ralph Lauren
(see L'Oreal Of Paris)

▼ Ponds
(see Chesebrough-Pond's)

💜 Porcelana Skin Care
(see Dep Corporation)

💜 Potions & Lotions/Body
& Soul ■ ● ✉
10201 North 21st Ave, Suite 8
Phoenix, AZ 85021
(800)456-3765, (602)944-6642
Personal care

♡ Power House Lime
Remover
(see SerVaas Labs)

▼ Power Stick
(see Chesebrough-Pond's)

▼ Pre Sun
(see Bristol-Myers Squibb)

💜 Precious Collection ☆ ■ ● ✉
P.O. Box 17155
Boulder, CO 80308
(303)447-1667
Personal care

▼ Prell
(see Procter & Gamble)

💜 Premier One Products ■
4133 South 89th Street
Omaha, NE 65127
(800)373-9660,
(402)339-9660
Personal care

💜 Prescriptives Inc.
(see Estee Lauder)

💜 Prestige Cosmetics ■ ●
1330 W. Newport Center Dr.
Deerfield Beach, FL 33442
Cosmetics, Personal care

💜 Prestige Fragrances, Ltd.
(see Revlon)

▼ Prince Matchabelli
(see Chesebrough-Ponds)

☆ Products contain no   ■ Manufacturer   ● Distributor   ✉ Mail Order
    animal or animal
    derived ingredients

💜 Princess Marcella
Borghese,Inc.
(see Revlon)

💜 Private Collection
(see Estee Lauder)

💜 Private Label ·
(see Twincraft)

💜 Pro-Care
(see Melaleuca)

💜 Pro-Gest
(see Professional &
Technical Services)

💜 Pro-Ma Systems
(USA) Inc. ☆ ●
477 Commerce Way, #113
Longwood, FL 32752-2109
(800)393-PROM,
(407)331-1133
*Cosmetics, Household,
Personal care*

💜 Pro-Tec Pet Health ●
P.O. Box 23676
Pleasant Hill, CA 94523
(800)44-FLEAS,
(510)676-9600
Companion animal

💜 ProAttitude
(see Neways, Inc.)

▼ Procter & Gamble Company
One Procter & Gamble Plaza
Cincinnati, OH 45201
(800)543-1745

💜 Professional & Technical
Services, Inc. ✉
621 SW Alder Street
Suite 900
Portland, OR 97205-3627
(800)648-8211
Personal care

▼ Professional Make-Up
Remover
(see Max Factor)

💜 Professional Pet
Products ☆ ■
1873 NW 97th Ave.
Miami, FL 33172
(800)432-5349,
(305)592-1992
Companion animal
Note: certain chemicals
contained in pesticide
products are required by law
to be tested on animals. All
other products are cruelty-free.

💜 CRUELTY FREE
Does not test products
or ingredients on animals

💜 Ingredients MAY
be tested on
animals

▼ Tests products or
ingredients
on animals

♡ Professional Supreme
Shampoo
(see House of Lowell)

▼ Prop Pre Shave
(see Mennen)

▼ Propert's Shoe Polish
(see Kiwi Brands)

▼ Protect
(see DowBrands, Inc.)

▼ Protein 21 Hair Groom
(see Mennen)

▼ Protein 29 Hair Groom
(see Mennen)

❤ Protein Plus
(see Palm Beach Beauty
Products)

▼ Protex
(see Colgate-Palmolive)

▼ Prouss' Mousse
(see Colgate-Palmolive)

❤ Psoria-Gard
(see Hobe Labs)

▼ Puffs
(see Procter & Gamble)

▼ Puffy Eye Minimizer
(see Max Factor)

❤ Pumice Stone
(see Delby System)

▼ Punch Laundry Detergent
(see Colgate-Palmolive)

❤ Puppetdears
(see Delby System)

❤ Pure & Natural
(see Dial Corporation)

❤ Pure Gold Aloe Vera ☆ ■ ●
P.O. Box 51867
Bowling Green, KY 42101
(502)796-8353
Personal care

❤ Pure Nature
(see Avanza)

❤ Pure Style
(see Levlad, Inc.)

💜 Purex
(see Dial Corporation)

▼ Purpose
(see Johnson & Johnson)

▼ Q-Tips Products
(see Chesebrough-Pond's)

💜 Q.C.
(see Key Distributors)

▼ QT
(see Schering-Plough)

♡ Qualis, Inc.
(see Bio Sentry Labs)

💜 Qualla
(see Palm Beach Beauty Products)

💜 Quan Yin Essentials ☆ ■
P.O. Box 2092
Healdsburg, CA 95448
(707)431-0529
Personal care

💜 Queen Helene
(see Para Labs)

▼ Quick Dip Silver Cleaner
(see Reckitt & Colman)

💜 Quick Tan
(see Palm Beach Beauty Products)

▼ Quiet Touch
(see Clairol)

💜 Quik Out
(see Magic American)

💜 Quinta Essentia
(see Norimoor)

💜 Quintessentials Flower Essence
(see Flower Essence Services)

💜 Rachel Perry, Inc. ● ✉
9111 Mason Avenue
Chatsworth, CA 91311
(800)966-8888,
(818)886-0202
Cosmetics, Personal care

💜 Radiant Glo
(see Palm Beach Beauty Products)

▼ Raid
(see S.C. Johnson & Son)

💜 Rain Forest Products
(see Green Mountain)

💜 Rainbow Light
(see Basically Natural)

♡ Rainbow Research Corp. ■ ✉
170 Wilbur Place
Bohemia, NY 11716
(800)722-9595, (516)589-5563
Personal care

♡ Rainbow Salon Botanicals
(see Rainbow Research)

💜 Rainforest
(see Green Mountain)

💜 Rainforest Essentials ☆ ■ ●
1718 22nd Street
Santa Monica, CA
90404-3921
(310)826-2143
Personal care

♡ Ralph Lauren ■
P.O. Box 98
Westfield, NJ 07091
(800)631-7358
Personal care

▼ Ranir/DCP Corporation
4701 East Paris Ave., SE
Grand Rapids, MI 49512
(616)698-8880

▼ Rave
(see Chesebrough Pond's)

💜 Ravenwood ■ ●
10 Timberlee Creek
Traverse City, MI 49684
(616)929-4181
Personal care

💜 RC International ☆ ■
11222 "I" Street
Omaha, NE 68137
(800)433-3970,
(402)592-2102
Personal care

💜 RCN Products, Inc. ☆ ■
92 Argonaut, Suite 275
Aliso Viejo, CA 92656
(714)581-6311
Household

▼ Reach
(see Johnson & Johnson)

💜 Real Aloe Company ■
P.O. Box 2770
Oxnard, CA 93033
(800)541-7809
Personal care

☆ Products contain no animal or animal derived ingredients ■ Manufacturer ● Distributor ✉ Mail Order

💜 Real Animal Friends ☆ ■
101 Albany Avenue
Freeport, NY 11520
(516)223-7600
Companion animal

💜 Real Salt
(see American Orsa)

💜 ReBalance
(see L'anza Research Int'l)

▼ Reckitt & Colman Inc.
1655 Valley Road
Wayne, NJ 07470
(201)633-6700

▼ Red Door
(see Elizabeth Arden)

💜 Redken Labs, Inc. ■
P.O. Box 832
Clark, NJ 07066-0832
(800)423-5369, (800)423-5280
Cosmetics, Personal care

💜 Redmond Clay
(see American Orsa)

♡ Redmond Products, Inc. ■
18930 West 78th Street
Chanhassen, MN 55317
(800)628-4094, (612)934-4868
Cosmetics, Personal care

💜 Reflections
(see R C Int'l)

💜 Regeneration
(see BeautiControl
Cosmetics)

💜 Regis
(see La Dove)

▼ Relax & 24 Hour
(see DowBrands, Inc.)

💜 Renascence Massage,
Bath & Body Oil
(see Pacific Scents)

💜 Rene Furlerer
(see LaNatura)

💜 Rene Guinot
(see Francosmetics Int'l)

▼ Renuzit
(see S.C. Johnson)

💜 Restor ☆ ■ ● ✉
7323 Beverly Blvd.
Los Angeles, CA 90036
(213)936-2191
*Personal care, Companion
animal*

❤ Reviva Labs, Inc. ■ ✉
705 Hopkins Road
Haddonfield, NJ 08033
(800)257-7774,
(609)428-3885
Cosmetics, Personal care

❤ Revlon, Inc. ■
625 Madison Avenue
New York, NY 10022
(800)4Revlon,
(212)572-5000
Cosmetics, Personal care

▼ Rhuli Products
(see S.C. Johnson)

▼ Right Guard
(see Gillette)

▼ Rinso
(see Lever Brothers)

❤ Rio Vista Marketing
Associates, Inc. ✩ ■
P.O. Box 60806
Santa Barbara, CA 93160
(800)248-6428
(805)968-9424
Companion animal

▼ Rise
(see Chesebrough Pond's)

❤ RJF, Inc./Oshadhi ✩ ●
15 Monarch Bay Plaza
Suite 346
Monarch Beach, CA 92629
(800)933-1008, (714)240-1104
Personal care

❤ RJG Inc. (Formerly
Dymer & Associates) ✩ ●
11243-4 St. John's
Industrial Parkway, S.
Jacksonville, FL 32246
(800)969-4591, (904)642-4591
Personal care

❤ Roach Pruff
(see Copper Brite)

❤ Roadside Flowers Odor
Eliminator
(see Earthly Matters)

❤ Rokeach Corporation ✩ ■
80 Avenue K
Newark, NJ 07105-3803
(201)587-1199
Household

▼ Rosewater
(see Max Factor)

❤ Ross The Boss
(see La Dove)

💜 Royal Herbal
(see Pet Connection)

💜 Royal Labs Natural
Cosmetics ☆ ■
87 Venice Avenue
Waterbury, CT 06708
(800)203-5151,
(203)753-2737
Cosmetics, Personal care

▼ Rug Fresh
(see Reckitt & Colman)

💜 Russ Kalvin's Hair Care ■
25655 Springbrook Ave., #6
Saugus, CA 91350
(805)253-2723
Cosmetics, Personal care

💜 Rx for Fleas, Inc. ☆ ■ ●
6555 NW 9th Avenue
Suite 412
Fort Lauderdale, FL 33309
(800)666-3532,
(305)351-9244
Companion animal

▼ S.C. Johnson & Son, Inc.
1525 Howe Street
Racine, WI 53403
(800)558-5252,
(414)631-2000

▼ S.O.S.
(see Clorox)

▼ S.O.S. Scouring Pads
(see Miles)

♡ Safari by Ralph Lauren
(see L'Oreal of Paris)

💜 Safechoice
(see AFM Enterprises)

💜 Safecoat
(see AFM Enterprises)

▼ Safeguard
(see Procter & Gamble)

💜 Safer Chemical Co. ☆ ■ ● ✉
4002 Minden Avenue
Texarkana, AR 75502
(501)773-4901
Household, Companion animal

💜 Safer Rug Shampoo
(see Safer Chemical)

💜 Sagami, Inc. ☆ ■ ●
825 N. Cass Avenue
Suite 101
Westmont, IL 60559
(708)789-9999
Personal care

♡ Sally Hansen
(see Del Laboratories, Inc.)

♥ Salon Nail Enamel Dryer
(see IQ Products)

♡ Salon Naturals
(see Shikai Products)

▼ Salon Selectives
(see Helene Curtis)

♡ Samsara
(see Guerlain, Inc.)

♥ San Francisco Soap
Company
(see Avalon Natural
Cosmetics)

♡ Sandahl
(see L.T. York)

▼ Sani-Flush
(see Reckitt & Colman)

▼ Sanofi Beaute, Inc.
40 East 52nd Street
New York, NY 10022-5911
(212)621-7300

♥ Santa Fe Fragrance
Inc. ☆ ■ ●
P.O. Box 282
Santa Fe, NM 87504
(505)473-1717
Personal care

♥ Sante Fe Soap
(see Basically Natural)

♥ Sappo Hill Soapworks ☆ ■
654 Tolman Creek Road
Ashland, OR 97520
(503)482-4485
Personal care

♥ Sarah Michaels ●
180 Campanelli Parkway
Stoughton, MA 02072
(617)341-8810
Personal care

▼ Satin
(see Max Factor)

♥ Saturday Nite Catnip
(see Pet Connection)

♥ Saurys Sport Sunblock
(see Finley Company)

♥ Sausalito Soaps
(see Clearly Natural)

☆ Products contain no   ■ Manufacturer   ● Distributor   ⊠ Mail Order
  animal or animal
  derived ingredients

💜 Scan Jet Flea Blaster
(see Copper Brite)

💜 Scarborough & Company
(see Crabtree & Evelyn, Ltd.)

♡ Scargo
(see Home Health Products)

💜 Scenational Blend Sampler
(see Pacific Scents)

💜 Schiff ■ ●
1911 South 3580 West
Salt Lake City, UT 84104
(800)526-6251, (801)972-0300
Personal care

💜 Schroeder & Tremayne,
Inc. ■ ●
1051 Cassens Industrial Ct.
Fenton, MO 63026
(800)325-3545,
(314)349-5655
Personal care

💜 Schwarzkopf, Inc. ■
5701 Buckingham Parkway
Suite E
Culver City,CA 90230
(310)641-4600
Personal care

▼ Scoop Fresh Scoopable
Cat Litter
(see Clorox)

▼ Scope
(see Procter & Gamble)

💜 Scott McClintock for Men
(see Jessica McClintock)

▼ Scott Paper Company
Scott Plaza
Philadelphia, PA 19113
(800)TEL-SCOT,
(215)552-5000

▼ Scott's Emulsion
(see SmithKline Beecham)

💜 Scoundrel
(see Revlon)

💜 Scruples, Inc. ■
8221-214th Street West
Lakeville, MN 55044-9102
(612)469-4646
Personal care

▼ Sea & Ski
(see Unilever U.S.)

💜 Sea Bath Body Vitalizer
(see Wachters')

💜 CRUELTY FREE
Does not test products
or ingredients on animals

♡ Ingredients MAY
be tested on
animals

▼ Tests products or
ingredients
on animals

▼ Sea Breeze
(see Bristol-Myers Squibb)

❤ Sea Caress Glycerine Soap
(see Wachters')

❤ Sea Minerals Co. ☆ ■
2886 Heath Ave.
Bronx, NY 10463
(718)796-5509
Cosmetics, Personal care

❤ Sea Spraa Plant &
Conditioning Products
(see Wachters')

❤ Sea Tone Facial Toner
(see Wachters')

❤ Sea Weed Masque
(see Wachters')

❤ Sea-Min Plant & Soil
Conditioning Products
(see Wachters')

❤ Seasonals
(see Sojourner Farms
Natural Pet Products)

♡ Sebastian International,
Inc. ☆ ■ ●
6109 DeSoto Avenue
Woodland Hills, CA 91367
(800)829-7322,
(818)999-5112
Cosmetics, Personal care

▼ Second Nature
(see Clairol)

▼ Secret
(see Procter & Gamble)

❤ Secret Garden Perfume Oils
(see Ananda Country
Products)

❤ Secret Products
(see Green Mountain)

❤ Secretary's Secret
(see Green Mountain)

❤ Seide Haircare Products
(see Borlind of Germany)

▼ Sensor
(see Gillette)

❤ Sensory Xperience
(see Pacific Scents)

☆ Products contain no
animal or animal
derived ingredients   ■ Manufacturer   ● Distributor   ⊠ Mail Order

💜 Sensuous Shadows with Silkenspheres
(see BeautiControl Cosmetics)

💜 Septi-Sea Astringent
(see Wachters')

💜 Serene
(see Melaleuca)

▼ Serenity Guards
(see Johnson & Johnson)

♡ SerVaas Laboratories Inc. ☆ ■
1200 Waterway Blvd.
Indianapolis, IN 46207
(800)433-5818,
(317)636-7760
Household

▼ Sesame Street Children's Bath Products
(see Colgate-Palmolive)

💜 Set-N-Me-Free Aloe Vera Company ☆ ■
19220 SE Stark
Portland, OR 97233-5751
(800)221-9727,
(503)666-9661
Cosmetics, Personal care, Companion animal

💜 Seventh Generation ● ✉
49 Hercules Drive
Colchester, VT 05446
(800)456-1177
Household, Personal care

▼ Shade Sunscreen
(see Schering-Plough)

▼ Shadow Perfume
(see Unilever U.S.)

💜 Shahin Soap ☆ ■ ✉
427 Van Dyke Ave.
P.O. Box 8117
Haledon, NJ 07508
(201)790-4296
Personal care

💜 Shaklee U.S., Inc. ■ ●
444 Market St.
San Francisco, CA 94111
(800)SHAKLEE,
(415)954-3000
Cosmetics, Personal care, Household

♡ Shalimar
(see Guerlain, Inc.)

💜 Shampoo with Panthenol
(see Wachters')

❤ Shatoiya's Handmade
(see Dry Creek Herb Farm)

▼ Shave Talc
(see Mennen)

❤ Shaver 2000
(see Bonne Bell)

❤ Sheer Protection
(see BeautiControl
Cosmetics)

▼ Shield
(see Lever Brothers)

♡ Shikai Products ■
P.O. Box 2866
Santa Rosa, CA 95405
(800)448-0298,
(707)584-0298
Personal care

▼ Shimmer Lights
(see Clairol)

❤ Shine Tech
(see BeautiControl
Cosmetics)

♡ Shiny Sinks
(see SerVaas Labs.)

❤ Shirley Price
Aromatherapy Ltd. ■ ✉
462 62nd Street
Brooklyn, NY 11220
(718)492-9514
Personal care

▼ Shiseido Cosmetics
(America) Ltd.
900 Third Avenue
New York, NY 10022
(212)752-2644
Cosmetics

▼ Shout
(see S.C. Johnson & Son)

▼ Shower to Shower
(see Johnson & Johnson)

❤ Siddha International ☆ ■
P.O. Box 5127
Gainesville, FL 32605
(904)376-8173
Personal care

❤ Sierra Dawn Products ☆ ■
P.O. Box 1203
Sebastopol, CA 95472
(707)577-0324
Personal care, *Household*

☆ Products contain no   ■ Manufacturer   ● Distributor   ✉ Mail Order
animal or animal
derived ingredients

💜 Sigma-Tau Consumer Health Prod. ☆ ■
200 Orchard Ridge Dr.
Suite 300
Gaithersburg, MD 20878
(800)447-0169,
(301)948-1041
Personal care

▼ Signal Mouthwash
(see Lever Brothers)

▼ Silkience
(see Gillette)

💜 Silver Fox
(see Carme)

💜 Simple Wisdom, Inc. ■ ✉
775 South Graham
Memphis, TN 38111
(800)370-6550, (901)458-4686
Personal care, *Household*

💜 Simplers Botanical Co. ■ ✉
P.O. Box 39
Forestville, CA 95436
(707)887-2012
Personal care, *Companion animal*

💜 Sinclair & Valentine
(see Smith & Vandiver)

💜 Skin
(see Bonne Bell)

▼ Skin Bracer
(see Mennen)

▼ Skin Care Products
(see Prince Matchabelli)

💜 Skin Saver
(see Palm Beach Beauty Products)

💜 Skin Savvy
(see Strong Skin Savvy)

♡ Skin Trip
(see Mountain Ocean)

💜 Skinlogics
(see BeautiControl Cosmetics)

▼ Skintimate
(see S.C. Johnson)

💜 Sleepy Hollow Botanicals
(see Carme)

💜 Slim Tea
(see Hobe Labs)

💜 Slopes Sunblock
(see Finley Company)

💜 CRUELTY FREE
Does not test products or ingredients on animals

💜 Ingredients MAY be tested on animals

▼ Tests products or ingredients on animals

💜 Smackers
(see Bonne Bell)

💜 Smith & Vandiver Inc. ■
480 Airport Boulevard
Watsonville, CA 95076
(408)722-9526
Cosmetics, Personal care

▼ SmithKline Beecham
Consumer Healthcare
P.O. Box 1467
Pittsburgh, PA 15230
(800)456-6670,
(412)928-1000

💜 Sno Drops
(see Dial Corporation)

💜 Sno-Bowl
(see Dial Corporation)

▼ Snuggle
(see Lever Brothers)

💜 Soap Factory ☆ ■ ●
141 Cushman Road
St. Catherines, Ontario
Canada L2M 6T2
(416)465-6691
*Household products,
Personal care*

💜 Soap Works ■ ●
60 Chatsworth Drive
Toronto, Ontario
Canada M4R 1R5
*Personal care, Companion
animal*

💜 SoapBerry Shop ■ ✉ ●
50 Galaxy Boulevard, #12
Rexdale, Ontario
Canada M9W 4Y5
(416)674-0248
Cosmetic, Personal care

💜 SoColor
(see Matrix Essentials)

💜 Sof' Touch
(see Jojoba Resources)

▼ Sof'Stroke
(see Mennen)

▼ Soft & Dri
(see Gillette)

💜 Soft 'N Fresh
(see Green Mountain)

▼ Soft Scrub
(see Clorox)

▼ Soft Sense
(see Bausch & Lomb)

☆ Products contain no
animal or animal
derived ingredients

■ Manufacturer ● Distributor ✉ Mail Order

💜 Softa Skin Hand &
Body Lotion
(see Wachters')

▼ Softsoap
(see Colgate-Palmolive)

▼ Softwash
(see Colgate-Palmolive)

💜 Soilove Laundry Soil &
Stain Remover
(see America's Finest
Products)

💜 Sojos
(see Sojourner Farms
Natural Pet Products)

💜 Sojourner Farms Natural
Pet Products ■
P.O. Box 8062
Ann Arbor, MI 48107
(800)767-6567,
(313)994-3974
Companion animal

▼ Solarcaine
(see Schering-Plough)

💜 Solarex Sunblock
(see Finley Company)

💜 Solid Gold Holistic Animal
Equine Nutrition Ctr. ■
1483 North Cuyamaca
El Cajon,CA 92020
(619)258-1914
Companion animal

▼ Solo
(see Procter & Gamble)

💜 Sombra Cosmetics Inc. ■ ✉
5600-G McLeod, NE
Albuquerque, NM 87109
(800)225-3963, (505)888-0288
Cosmetics, Personal care

▼ Some Color Mascara
(see Max Factor)

💜 Song of Life, Inc. ■
152 Fayette St.
Buckhannon, WV 26201
(304)472-6114
Personal care

▼ Soupline
(see Colgate-Palmolive)

💜 Spanish Bath
(see San Francisco Soap
Company)

💜 Sparkle Glass Cleaner
(see A.J. Funk)

💜 CRUELTY FREE
Does not test products
or ingredients on animals

💜 Ingredients MAY
be tested on
animals

▼ Tests products or
ingredients
on animals

❤ Sparkle Paper Towels
(see Georgia Pacific Corp.)

❤ Spectaculash
(see BeautiControl
Cosmetics)

▼ Speed Stick
(see Mennen)

❤ Spellbound
(see Estee Lauder)

▼ Spic & Span
(see Procter & Gamble)

❤ Spirit
(see Dial Corpoation)

❤ Spirit of Saint Alban ☆ ■ ●
441 Little Elk Creek Road
Elkton, MD 21921
(800)453-3883, (410)392-0253
Cosmetics

❤ Sponge Drop Wedges
(see Delby System)

▼ Sport Strip
(see Johnson & Johnson)

❤ Sports Nutrition
(see C.E. Jamieson)

❤ Spra-itt
(see Wachters')

▼ Spray 'n Vac
(see Reckitt & Colman)

▼ Spray 'N Wash
(see DowBrands)

❤ Spray Zapper
(see Alpha 9)

▼ Springfield Products
2601 S. Eastern Avenue
Los Angeles, CA 90040
(213)723-7476

❤ St. Clair Industries, Inc. ☆ ●
3067 E. Commerical Blvd.
Ft. Lauderdale, FL 33308
(305)491-0401
*Personal care, Companion
animal*

❤ St. John's Herb Garden,
Inc. ☆ ■ ● ✉
7711 Hillmeade Road
Bowie, MD 20720
Personal care, Household

❤ Sta-Flo
(see Dial Corporation)

💜 Sta-Puf
(see Dial Corporation)

▼ Stain Out
(see Clorox)

▼ Stain Stick
(see DowBrands)

▼ Stanley Home Products
50 Payson Avenue
Easthampton, MA
01027-2258
(413)527-1000

💜 Star Brite ☆ ■
4041 S.W. 47th Avenue
Fort Lauderdale, FL 33314
(800)327-8583,
(305)587-6280
Household

💜 Star of the East Incense
(see Ananda Country
Products)

💜 Starlet O'Hara
(see Delby System)

💜 Starry Nights Glass Cleaner
(see Earthly Matters)

💜 Starwest Botanicals, Inc. ☆ ■
11253 Trade Center Drive
Rancho Cordova, CA 95742
(800)800-4372,
(916)638-8100
Personal care

▼ Static Guard
(see Alberto-Culver)

💜 Stature Field
Corporation ☆ ● ✉
1143 Rockingham Drive
Suite 106
Richardson, TX 75080
(214)504-8845
Personal care

▼ Stayfree
(see Johnson & Johnson)

▼ Stendhal
(see Sanofi Beaute)

▼ Step Saver
(see S.C. Johnson & Son)

💜 Steps In Health, Ltd. ✉
P.O. Box 1409
Lake Grove, NY 11755
(800)471-VEGE,
(516)471-2432
Cosmetics, Personal care

▼ Stick-Ups
(see Reckitt & Colman)

♡ Stick-With-Us
Products, Inc. ☆ ■
P.O. Box 26190
Richmond, BC
Canada V6Y 3V3
(604)241-0448
Cosmetics

❤ Stiefel Laboratories, Inc. ■
255 Alhambra Circle
Coral Gables, FL 33134
(800)327-3858,
(305)443-3800
Personal care

❤ Stonybrook Botanicals
(see Rainbow Research)

❤ Stress Relief
(see Palm Beach Beauty
Products)

❤ Stress Therapy
(see Palm Beach Beauty
Products)

▼ Stretch Mascara
(see Max Factor)

♡ Strickland, J. & Company ■ ●
P.O. Box 840
Memphis, TN 38101
Cosmetics, Personal care

❤ Strong Skin Savvy, Inc. ■
1 Lakeside Drive
New Providence, PA 17560
(800)724-3952,
(717)786-8947
Personal care

♡ Studio Line
(see L'Oreal of Paris)

❤ Studio Magic Inc. ☆ ●
1417-3 Del Prado Blvd., #480
Cape Coral, FL 33990-3749
(813)283-5000
Cosmetics

▼ Style
(see DowBrands)

▼ Style Plus
(see DowBrands)

▼ Suave
(see Helene Curtis)

▼ Suavitel
(see Colgate-Palmolive)

💜 Success Cologne
(see BeautiControl
Cosmetics)

▼ Sudden Tan
(see Schering-Plough)

💜 Suisse Spa Suisse
(see C.E. Jamieson)

💜 Sukesha
(see Chuckles)

💜 Sumeru Garden Herbals ☆ ■
P.O. Box 2110
Freedom, CA 95019
(408)722-4104
Personal care

▼ Summer Blonde
(see Clairol)

▼ Summer Gold
(see Marche Image)

▼ Summit
(see Procter & Gamble)

💜 Sun Defiance
(see Palm Beach
Beauty Products)

💜 Sun Detergent
(see Huish Detergents)

♡ Sun Pharmaceuticals Ltd. ■ ●
700 Fairfield Ave.
Stamford, CT 06904
Personal care

💜 Sun Shades
(see Melaleuca)

▼ Sundown
(see Johnson & Johnson)

💜 Sunfeather Herbal Soap
Company ☆ ■ ✉
HCR 84-Box 60a
Potsdam, NY 13676
(315) 265-3648
*Personal care, Companion
animal*

💜 Sunflowers Dishwashing
Liquid
(see Earthly Matters)

💜 Sunless Bronze
(see Borlind of Germany)

▼ Sunlight
(see Lever Brothers)

💜 Sunlind Sun Cosmetics
& Gels
(see Borlind of Germany)

💜 CRUELTY FREE
Does not test products
or ingredients on animals

💜 Ingredients MAY
be tested on
animals

▼ Tests products or
ingredients
on animals

💜 Sunlogics
(see BeautiControl
Cosmetics)

💜 Sunrise Lane Products, Inc. ✉
780 Greenwich Street
New York, NY 10014
(212)242-7014
Cosmetics, Personal care,
Household

💜 Sunshine Flowers &
Fragrances ☆ ■
1616 Preuss Rd.
Los Angeles, CA 90035
(310)5275-9891
Cosmetics, Personal care

💜 Sunshine Natural
Products ☆ ■
Route 5W
Renick, WV 24966
(304)497-3163
Cosmetics, Personal care,
Companion animal

▼ Super Blue
(see Gillette)

💜 Super Dry Industries ☆ ■
P.O. Box 49
Highway 39 South
Shuqualak, MS 39361
(800)808-CATS,
(601)793-9555
Companion animal

💜 Super Pine
(see Green Mountain)

▼ Super Suds
(see Colgate-Palmolive)

▼ SuperBait
(see Clorox)

▼ Supreme au Creme
(see Lever Brothers)

💜 Surco Products, Inc. ☆ ■
P.O. Box 777
Braddock, PA 15104
(412)351-7700
Personal care, Household

▼ Sure
(see Procter & Gamble)

▼ Sure & Natural
(see Johnson & Johnson)

☆ Products contain no   ■ Manufacturer   ● Distributor   ✉ Mail Order
   animal or animal
   derived ingredients

▼ Surefit
(see Johnson & Johnson)

▼ Surevue
(see Johnson & Johnson)

▼ Surf
(see Lever Brothers)

♥ Susan Lucci Hair Care ☆
505 S. Beverly Drive
Suite 683
Beverly Hills, CA 90210
Personal care

♥ Swasthya
(see Auromere Ayurvedic
Imports)

♥ Swedish Pollenique
(see Cernitin America)

♥ Swedish Supreme
(see Cernitin America)

♥ Sweetheart
(see Dial Corporation)

▼ Swish Toilet Bowl Cleaner
(see Reckitt & Colman)

▼ Swivel
(see Gillette)

♥ Synerfusion
(see Matrix Essentials)

♥ Systeme Biolage
(see Matrix Essentials)

♥ T'rific
(see Wellington Labs)

▼ Tackle
(see Clorox)

♥ Talisman ● ✉
68 Tinker Street
Woodstock, NY 12498
(914)679-7647
Personal care

♡ Tambrands Inc. ☆ ■
777 Westchester Ave.
White Plains, NY 10604
(800)523-0014,
(914)696-6770
Personal care

▼ Tame
(see Gillette)

♡ Tampax Tampons
(see Tambrands)

▼ Tan Accelerator
(see Unilever U.S.)

❤ Tana
(see Kiwi Brands)

❤ Tau-Marin Toothpaste Gel
(see Sigma-Tau Consumer
Health Products)

❤ TAUT by Leonard
Engelman ■
9428 Eton #M
Chatsworth, CA 91311
(800)438-8288
Cosmetics, Personal care

❤ TCCD International,
Inc. ☆ ■ ●
3012 NW 25th Avenue
Pompano Beach, FL 33069
(800)653-4006,
(305)960-4904
Cosmetics, Personal care

❤ Technology Flavors and
Fragrances ■
10 Edison Street E.
Amityville, NY 11701
(516)842-7600
Personal care

▼ Tempra
(see Bristol-Myers Squibb)

❤ Ten
(see E. Burnham Cosmetics)

❤ Ten-O-Six Lotion
(see Bonne Bell)

▼ Tenax Hair Styling
(see Chesebrough-Pond's)

❤ Tender Age
(see C.E. Jamieson & Co. Ltd.)

♡ Tender Corporation ☆ ■
P.O. Box 290
Littleton Industrial Park
Littleton, NH 03561
(603)444-5464
Household
Note: certain chemicals
contained in insect repellants
are required by law to be
tested on animals.

❤ Terme di Montecatini
(see Revlon)

❤ Terra Flora Essential Oils
(see Flower Essence Services)

❤ Terra Flora Herbal Body
Care Products ☆ ■ ● ⊠
P.O. Box 177
Montague, MA 01351
(413)367-0365
Personal care

☆ Products contain no animal or animal derived ingredients   ■ Manufacturer   ● Distributor   ⊠ Mail Order

♡ Terracotta
(see Guerlain, Inc.)

♥ TerraNova
(see AKA Saunders, Inc.)

♥ Terressentials ☆ ■ ✉
3320 N. 3rd Street
Arlington, VA 22201-1712
(703)525-0585
Household, Personal care,
Companion animal

♥ Terry Laboratories, Inc. ☆ ■
390 North Wickham Rd.
Suite F
Melbourne, FL 32935
(800)367-2563, (407)259-1630
Personal care

♥ Texmaco USA
(see Goodebodies)

♥ Thai Deodorant Stones
(see Deodorant Stones of
America)

♥ Thai Stick & Pure &
Natural
(see Deodorant Stones of
America)

♥ That Man
(see Revlon)

♥ Therapeutic Hand Lotion
(see E. Burnham Cosmetics)

♥ Therapeutic-Dermato-
logic Formula
(see Dermatologic
Cosmetics Labs)

▼ Thick Again
(see Max Factor)

♥ Third Millennium
Science ☆ ■
2195 Faraday Avenue, #F
Carlsbad, CA 92008
(800)776-6525, (619)431-7181
Personal care, Household,
Companion animal

♥ Thursday Plantation
Labs, Pty., Ltd. ■ ● ✉
P.O. Box 5613
Montecito, CA 93150-4613
(805)963-2297
Personal care, Household,
Companion animal

▼ Tickle
(see Bristol-Myers Squibb)

▼ Tide
(see Procter & Gamble)

♥ CRUELTY FREE
Does not test products
or ingredients on animals

♥ Ingredients MAY
be tested on
animals

▼ Tests products or
ingredients
on animals

❤ Tiffany & Company ☆ ■
727 Fifth Avenue
New York, NY 10022
(212)755-8000
Personal care

❤ Tigi
(see La Dove)

▼ Tigress
(see Unilever U.S.)

▼ Tilex
(see Clorox)

♡ Timberline
(see Mem Company)

♡ Tinkerbell
(see Mem Company)

▼ Tintex Fabric Dye
(see Kiwi Brands)

❤ Tisserand Aromatherapy ☆ ■
(see Avalon Natural
Cosmetics)

❤ TN Dickinson ☆ ■
313 E. High Street
East Hampton, CT 06424
(800)203-4444
Cosmetics, Personal care

❤ Tom Fields. Ltd.
(see Mem Company, Inc.)

❤ Tom's of Maine ■
P.O. Box 710
Kennebunk, ME 04043
(800)367-8667, (207)985-2994
Personal care

❤ Tone
(see Dial Corporation)

▼ Toni
(see Gillette)

▼ Top Job
(see Procter & Gamble)

❤ Topol Toothpaste
(see Dep Corporation)

▼ Torrids
(see Clairol)

❤ Total Shaving Solution ☆ ■
P.O. Box 832074
Richardson, TX 75083
Personal care

▼ Touche Bath Oil
(see DowBrands, Inc.)

▼ Toujours Moi
(see Max Factor)

144

♡ TR3
(see Blue Coral)

♥ Travelette
(see Delby System)

▼ Trac II
(see Gillette)

♥ Trend
(see Dial Corporation)

♥ Trader Joe's Company ■ ●
538 Mission Street
South Pasadena, CA 91030
(818)441-1177
Personal care, Household

▼ Tresemme
(see Alberto-Culver)

♥ Tressa, Inc. ■
P.O. Box 75320
Cincinnati, OH 45275
(800)879-8737,
(606)525-1300
Personal care

♥ Traditional Flower Remedies
(see Ellon USA)

♥ Traditional Products ■
P.O. Box 564
Crewell, OR 97426
(503)895-2957
Personal care, Household

♡ Trevor Sorbie International ■
8033 Sunset Blvd., Ste 6226
Los Angeles, CA 90046
(213)656-4499
Personal care

♥ Tranquil Waters Body Wash
(see Earthly Matters)

♡ Trewax
(see Blue Coral)

♡ Trans-India(Shikai) ■
Box 2866
Santa Rosa, CA 95405
Personal care

♥ Tri
(see Institute of Trichology)

♥ Travel Mates America ■
1760 Lakeview Road
Cleveland, OH 44112
(216)231-4102
Personal care

♥ Trianco-Babycakes ■ ✉
14 Buchanan Road
Salem, MA 01970
(800)972-3364
Personal care

♥ CRUELTY FREE
Does not test products
or ingredients on animals

♥ Ingredients MAY
be tested on
animals

▼ Tests products or
ingredients
on animals

❤ Tropical Balm
(see Aurora Henna)

▼ Tropical Blend
(see Schering-Plough)

❤ Tropical Soap Company ■
P.O. Box 797217
Dallas, TX 75379
(800)527-2368, (214)243-1991
Personal care

❤ Tropix Actifier
(see Tropix Suncare)

❤ Tropix Suncare Products ■
217 S. 7th St., Ste 104
Brainerd, MN 56401
(800)421-7314
Cosmetics

❤ Truly Moist
(see Desert Naturels)

❤ TSP Powered All
Purpose Cleaner
(see America's Finest
Products Corp.)

▼ Tube Shave Cream
(see Mennen)

▼ Tuffy
(see Clorox)

❤ Tuscany
(see Estee Lauder)

▼ Twice As Fresh
(see Clorox)

▼ Twinkle
(see S.C. Johnson & Son, Inc.)

♡ Twinscents ●
2 Tigan St.
Winooski, VT 05404
(800)792-7377, (802)655-2200
Personal care

▼ Two Thousand (2000)
Calorie Mascara
(see Max Factor)

▼ Ty-D-Bol
(see Kiwi Brands)

❤ U.S. Sales Service ☆ ■
1414 E. Libra Dr.
Tempe, AZ 85283
(800)487-2633,
(602)839-3761
Personal care

❤ Ultima II
(see Revlon)

▼ Ultra Brite
(see Colgate-Palmolive)

☆ Products contain no animal or animal derived ingredients   ■ Manufacturer   ● Distributor   ⊠ Mail Order

💜 Ultra Glide Blue Strip
(see Wilkinson Sword)

💜 Ultra Glow Cosmetics ☆ ■ ●
P.O. Box 1469, Station A
Vancouver, BC
Canada V6C 1P7
(604)444-4099
Cosmetics

💜 Ultra Laundry
(see Earth Friendly
Products)

💜 Ultra Liquid Laundry
Detergent
(see Venus Labs)

💜 Ultra-Fresh
(see Palm Beach
Beauty Products)

💜 UltraSlim Tea
(see Hobe Labs)

▼ Ultress
(see Clairol)

💜 Un-Petroleum Jelly
(see Autumn Harp)

💜 Un-Soap
(see Action Labs)

💜 Unbelievable Blush
(see BeautiControl
Cosmetics)

💜 Uncommon Scents Inc. ■ ● ✉
P.O. Box 1941
Eugene, OR 97440
(800)426-4336,
(503)345-0952
Personal care

💜 Unelko Corporation ☆ ■
7428 E. Karen Drive
Scottsdale, AZ 85260
(602)991-7272
Household

💜 Uni-Kleen
(see Wachters')

💜 Unicure ■
4385 International Blvd.
Norcross, GA 30093
(404)925-4811
Personal care

▼ Unilever United States Inc.
390 Park Avenue
New York, NY 10022
(800)451-6679,
(212)888-1260

❤ Universal Light ☆ ●
P.O. Box 261
Wilmot, WI 53192-0261
(414)889-8571
Personal care

❤ Uplifting Energizer
(see Pacific Scents)

❤ USA Labs
(see Paul Mazzotta, Inc.)

❤ Val-Chem Company Inc. ☆ ■
P.O. Box 330
Sayre, PA 18840
(717)888-2205
Household

❤ Value Plus
(see Conair)

▼ Van Cleef & Arpels
(see Sanofi Beaute)

❤ Vanda Beauty Counselor ●
P.O. Box 3433
Orlando, FL 32802
(407)839-0223
Cosmetics, Personal care

♡ Vanderbilt by Gloria
Vanderbilt
(see L'Oreal of Paris)

▼ Vanish
(see S.C. Johnson & Son)

♡ Vapor Products ☆ ■
P.O. Box 56839
Orlando, FL 32586-8395
(800)621-2943, (407)851-6230
Household

▼ Vaseline
(see Chesebrough-Pond's)

❤ Vavoom
(see Matrix Essentials)

❤ Vegan Market ☆ ✉
8512 12th Avenue, NW
Seattle, WA 98117
(206)789-2016
*Cosmetics, Personal care,
Household, Companion animal*

❤ Vegelatum
(see Green Mountain)

▼ Vel Beauty Bar
(see Colgate-Palmolive)

❤ Velva Sea Clenzing Creme
(see Wachters')

♡ Velvet-Peach Hand
Cream & Lotion
(see House of Lowell)

☆ Products contain no
animal or animal
derived ingredients
■ Manufacturer   ● Distributor   ✉ Mail Order

▼ Venezia
(see Procter & Gamble)

💜 Venus & Apollo ✉
P.O. Box 436
New Market, NH
03857-0436
(603)659-2938
Cosmetics, Personal care

💜 Venus Laboratories, Inc. ☆ ■
855 Lively Blvd.
Wood Dale, IL 60191
(800)592-1900,
(708)595-1900
Household, Personal care

💜 Vermont Country Soap ☆ ■
76 Exchange Street
Middlebury, VT 05753-1105
(802)247-3357
Personal care

▼ Verve
(see Chesebrough-Pond's)

💜 Vi-Jon Laboratories, Inc. ☆ ■
6300 Etzel Avenue
St. Louis, MO 63133-1997
(800)325-8167
Personal care

💜 Vi-Rid
(see Friendly Systems)

▼ Vibrance
(see Helene Curtis)

💜 Vicco Toothpaste
(see Basically Natural)

♡ Victor of Milano, Ltd.
(see Mem Company)

💜 Victora's Secret
(see Gryphon Development)

💜 Victoria Jackson
Cosmetics, Inc. ☆ ✉
16 Poali Corporate Center
Paoli, PA 19301
(800)848-7990
Cosmetics, Personal care

▼ Vidal Sassoon
(see Procter & Gamble)

▼ Village Bath Products
(see Colgate-Palmolive)

💜 Viper Cleaner
(see Minto Industries)

💜 Virginia's Soap Limited ■
Group 60 Box 20, RR #1
Anola, Manitoba
Canada R0E 0A0
(204)866-3788
Personal care

💜 Vista Detergent
(see Huish Detergents)

💜 Vita-Fusion
(see Palm Beach Beauty
Products)

💜 Vita-Glo
(see Action Labs)

💜 Vita-Guard
(see Pro-Tec Pet Health)

▼ Vitale
(see DowBrands)

💜 Vital 21
(see Pet Connection)

💜 Vital Care
(see Key Distributors)

▼ Vital Eyes
(see SmithKline Beecham)

💜 Vital Health Products
Ltd. ☆ ■ ✉ ●
P.O. Box 164
Muskego, WI 53150
(414) 679-1846
Personal care

▼ Vitalis
(see Clairol)

▼ Vivid
(see DowBrands)

▼ VO5
(see Alberto-Culver)

💜 Wachters' Organic Sea
Products Corp. ☆ ■ ●
360 Shaw Road
South San Francisco, CA
94080
In CA (800)682-7100,
Outside CA (800)822-6565
*Personal care, Household,
Companion animal*

▼ Walgreen Company
(see Xcel Labs)

💜 WalkAbout
(see Thursday Plantation
Labs)

▼ Wall Power
(see S.C. Johnson & Son)

♡ Warm Earth
Cosmetics ■ ● ✉
1155 Stanley Avenue
Chico, CA 95928-6944
(916)895-0455
Cosmetics

☆ Products contain no   ■ Manufacturer ● Distributor ✉ Mail Order
animal or animal
derived ingredients

♥ Warm Skin
(see Aurora Henna)

♥ Warren Laboratories, Inc.
12603 Executive Drive
Stafford, TX 77477
(713)240-2563
Cosmetics, Personal care

♡ Wash N Go
(see Home Health Products)

▼ Waterman
(see Gillette)

▼ Waterproof Cream
Make-Up
(see Max Factor)

♡ Wedge Energy Products
(see Hysan Corp.)

♥ Weleda, Inc. ■ ✉
P.O. Box 249
Congers, NY 10920
(914)268-0406, (914)268-8572
*Personal care, Companion
animal*

♥ Wella Corporation ■ ●
524 Grand Ave.
Englewood, NJ 07631
(201)569-1020
Personal care

♥ Wellbody
(see Key Distributors)

♥ Wellington Laboratories,
Inc. ☆ ●
2488 Townsgate Road
Unit C
West Lake Village, CA 91361
(800)835-8118, (805)495-2455
Personal care

♥ Western Fields Floor Cleaner
(see Earthly Matters)

♡ Westley's
(see Blue Coral)

♥ Whip-It Products, Inc. ☆ ■ ●
P.O. Box 30128
Pensacola, FL 32503
(904)436-2125
Household

▼ White Diamonds
(see Elizabeth Arden)

♥ White Flower Analgesic
Balm
(see Janta Int'l)

♥ White King Detergent
(see Huish Detergents)

♥ CRUELTY FREE
Does not test products
or ingredients on animals

♥ Ingredients MAY
be tested on
animals

▼ Tests products or
ingredients
on animals

❤ White Linen
(see Estee Lauder)

▼ White Rain
(see Gillette)

▼ White Shoulders
(see Unilever U.S.)

❤ Whole Spectrum Beauty
& Wellness Products Corp.
(see Essential Products
of America)

▼ Wild Rain
(see Gillette)

❤ Wildfire Bowl Cleaner
(see Earthly Matters)

❤ Wilkinson Sword Inc. ☆ ■
7012 Best Friend Road
Atlanta, GA 30340
Personal care, *Household*

▼ Williams Shave Products
(see SmithKline Beecham)

▼ Wind Drift
(see Mennen)

▼ Wind Song
(see Chesebrough-Pond's)

▼ Windex
(see S.C. Johnson & Son)

❤ Windrose Trading
Company ☆ ●
P.O. Box 990
634 Schoolhouse Road
Madison, VA 22727
Personal care

❤ Winter Tan
(see Palm Beach Beauty
Products)

❤ Winter White
(see Green Mountain)

❤ Wirth International ☆ ■
20000 National Avenue
Hayward, CA 94545
(510)785-2505
Household

❤ WiseWays Herbals ■
Singing Brook Farm
99 Harvey Road
Worthington, MA 01098
(413)238-4268
Personal care

▼ Wisk
(see Lever Brothers)

☆ Products contain no
animal or animal
derived ingredients

■ Manufacturer   ● Distributor   ⊠ Mail Order

▼ Wizard Air Freshener
(see Reckitt & Colman)

▼ Women's Fragrance
(see Chesebrough-Pond's)

▼ Wondergrip
(see Johnson & Johnson)

▼ Wood Preen Floor Wax
(see Kiwi Brands)

▼ Woodhue
(see Chesebrough-Pond's)

▼ Woodrich
(see S.C. Johnson & Son)

▼ Woolite
(see Reckitt & Colman)

💜 World's Greatest
(see Avanza)

💜 Wysong Corporation ■
1880 North Eastman Road
Midland, MI 48640
(800)748-0188,
(517)631-0009
Personal care, Companion animal

▼ X-Rated Lip Gloss
(see Max Factor)

▼ Xcel Laboratories
188 Industrial Drive
Elmhurst, IL 60126
(708)279-7887

💜 Xia Xiang
(see Revlon)

▼ Yes
(see DowBrands)

💜 Yi'Ang Analgesic Balm
(see Wachters')

♡ York
(see L.T. York)

💜 Yves Rocher, Inc. ☆ ■ ✉ ●
1305 Goshen Parkway
West Chester, PA 19380
(800)321-YVES,
(610)430-8200
Cosmetics, Personal care

♡ Yves Saint Laurent ■ ●
40 West 57th Street
New York, NY 10019
(212)621-7300
Personal care

💜 Zand Herbal Formulas
(see McZand Herbals)

❤ Zenith Advanced Health
Systems International,
Inc. ■ ● ✉
P.O. Box 1739
Corvallis, OR 97339
(800)547-2741,
(503)754-7380
Household, Personal care

▼ Zest
(see Procter & Gamble)

❤ Zia Cosmetics ■ ✉
410 Townsend St., 2nd Flr.
San Francisco, CA 94107
(800)334-7546, (415)543-7546
Cosmetics, Personal care

❤ Zinc and Aloe
(see Aloe Up, Inc.)

▼ Zotos International, Inc.
100 Tokeneke Road
Darien, CT 06820-1005
(800)242-WAVE,
(203)655-8911

▼ Zud Cleanser
(see Reckitt & Colman)

COMPANIES THAT DID NOT RESPOND TO THE NAVS QUESTIONNAIRE

3M Home & Commercial
Care Products (ScottBrite)
P.O. Box 33068
St. Paul, MN 55133-33068

A&M Products
P.O. Box 1999
Danbury, CT 06813-1999
(800)444-MEOW

Abbott Laboratories
1 Abbott Park Road
Abbott Park, IL 60064-3500
(708)937-6100

Advance Laboratories
120 Elm Street
Watertown, MA 02172

African Bio-Botanica
1511 NW 6th St.
Gainesville, FL 32601-4019

AH Robbins Co.,
Consumer Products Div.
P.O. Box 26609
Richmond, VA 23261-6609

Alicia Karpati, Inc.
3570 South Ocean Blvd.
Palm Beach, FL 33480

Alleghany Pharmacal Corp.
277 N. Blvd.
Great Neck, NY 11022
(516)466-0660

AllerCare, Inc.
1660 Highway 100
South, #340
Minneapolis, MN
55416-1531

Allergan
2525 Dupont Drive
Irvine, CA 92713
(800)347-5005

Aloe USA
10117 W. Oakland Park
Blvd., #377
Fort Lauderdale, FL
33351-6917

Aloe Vera International
12901 Nicholson, Ste. #370
Farmers Branch, TX
75234

Alpine Aromatics
International Inc.
51 Ethel Rd., West
Piscataway, NJ 08854-
1348

Alvera Deodorant Fine
Industries
P.O. Box 1876
Kerrville, TX 78021

American Colloid
Company
South Railroad Street
Mounds, IL 62964
(618)745-9527

American Home Products
5 Giralda Farms
Madison, NJ 07940
(201)660-5000

American Naturals, Inc.
131 NW 4th Ste. #103
Corvalis, OR 97330

American Safety
Razor Co.
P.O. Box 500
Staunton, VA 24401

Amole' Inc.
2425 West Dorothy
Lane
Dayton, OH 45439
(513)294-0571

Amoresse Laboratories
4121 Tigris Way
Riverside, CA 92503
(800)258-7931

An-Tech Research Labs
201 N. Figueroa, Ste. 1487
Los Angeles, CA 90012
(213)975-1487

Anne Klein II
530 7th Ave.,18th Fl.
New York, NY 10018

Aretae Products, Inc.
18930 W. 78th Street
Minneapolis, MN 55317

Artmatic Cosmetics
4014 First Ave
Brooklyn, NY 11232

Aura Industries
6352 N. Lincoln Avenue
Chicago, IL 60659-1213
(312)588-8722

Ausimont USA, Inc.
44 Whippany Rd
Morristown, NJ 07960-
1838

Ayagutaq
Box 237
Hyampom, CA 96046-0237

Back to Nature, Inc.
5627 North Milwaukee
Chicago, IL 60646

Banite, Inc.
47· E. Market St.
Buffalo, NY 14204

Barcolene Company
P.O. Box R
Holbrook, MA 02343

Barone Cosmetics
200 S. Venice Blvd.
Venice, CA 90291-4537

Barsamian's
1030 Massachusetts Ave.
Cambridge, MA 02118

Basch Company, Inc.,
P.O. Box 188
Freeport, NY 11520

Beauty Development
Corporation
P.O. Box 2260
Delray Beach, FL 33447

Beauty Labs, Inc.
60 Oser Avenue
Hauppauge, NY 11788
(516)273-5100

Bee Creek Botanicals
P.O. Box 204056
Austin, TX 78720-4056

Ben Rickert, Inc.
359 Newark Pompton
Turnpike
Wayne, NJ 07470

Benckiser Consumer
Products
55 Federal Road
Danbury, CT 06813
(203)731-5000

Benefit Cosmetics
333 Kearny Street
Suite 200
San Francisco, CA 94108
(800)781-2336

Beverly Hills Cold Wax
P.O. Box 600476
San Diego, CA 92160

Bio Dynamax
6565 Odell Place
Boulder, CO 80301-3330

Bio Pure
770 Forest, #B
Birmingham, MI 48009

Bio-San Laboratories
P.O. Box 325
Derry, NH 03038
(800)848-2542

Blistex
1800 Swift Dr
Oak Brook, IL 60521-1501
(708)571-2870

Block Drug Company
1 Fawcett Drive
Del Rio, TX 78840

Blue Rhubarb, Inc.
Old Creamery Rd.
Harmony, CA 93435
(800)926-1017

Body Glove Skin &
Hair Care
406 Amapola Ave.
Suite 105
Torrance, CA 90501
(310)320-5550

Body Gold
5930 La Jolla Hermosa,
Dept. NW-04
La Jolla, CA 92037
(800)808-8448,
(619)459-2661

Bodycology
1845 W. 205th Street
Torrance, CA 90501
(310)328-9610

Boni New Derm Labs
815 Colorado Blvd.
Los Angeles, CA 90041

Bonjour Parfums, Inc.
65 West 55th Street,
Suite 303
New York, NY 10019
(212)582-4568

Botanical Laboratories,
Inc.
1441 West Smith Rd.
Bellingham, WA 98226
(800)232-4005

Boyle-Midway
Household Products, Inc.
1655 Valley Road
Wayne, NJ 07470
(201)633-6700

Bricker Labs, Inc.
18722 Santee Lane
Valley Center, CA 92082
(800)952-9568,
(619)749-7053

Bristol Marketing
4 Old Chimney Rd.
Barrington, RI 02806
(401)245-7030

Bronner Brothers
600 Bronner Brothers Way
Atlanta, GA 30310
(404)577-4321

Butler Company
4635 W. Foster Ave
Chicago, IL 60630
(312)777-4000

Buty Wave Products
Co., Inc.
7323 Beverly Blvd.
Los Angeles, CA 90036

California Cosmetics, Inc.
P.O. Box 9890
Calabasas, CA 91372
(818)341-6675

California Mango
16632 Burke Lane
Huntington Beach, CA
92647
(714)375-2599

Calvin Klein
Cosmetics Company
Trump Tower
725 Fifth Ave.
New York, NY 10022-
2519

Cameo, Inc.
P.O. Box 535
Toledo, OH 43693

Cantol, Inc.
2211 North America St.
Philadelphia, PA 19133

Caraloe, Inc.
2001 Walnut Hill Lane
Irving, TX 75038
(800)358-2588

Cardinal Laboratories, Inc.
710 Ayon Avenue
Azusa, CA 91702
(818)969-3305

Carewell
P.O. Box 810156
Boca Raton, FL 33481

Carson Products Co.
P.O. Box 22309
Savannah, GA 31403
(912)651-3400

Carter's Naturals
3A Hamilton Business
Park, P.O. Box 311
Dover, NJ 07801
(201)989-8880

Carter-Wallace
1345 Avenue of the
Americas
New York, NY 10105
(212)339-5000

CCA Industries, Inc.
200 Murray Hill Parkway
East Rutherford, NJ 07073
(201)935-3232

Cell Tech
1300 Main Street
Klamath, OR 97601-5914

Cellife International
1185 Linda Vista Dr.
San Marcos, CA 92069

Chamberlain Health &
Beauty Aids Company
P.O. Box 3570
Des Moines, IA 50322

Chattem Inc.
Lot 715 W. 38th Street
Chattanooga, TN 37409
(615)821-4571

Chemco Industries, Inc.
500 Citadel Drive
Suite 120
Los Angeles, CA 90040
(213)721-8300

Cher Beauty Products
208 Carter Dr.
West Chester, PA 19382

Chico-San
Cosmetics & Toiletries
P.O. Box 1050
Gridley, CA 95948

Chinese Native Products
393 W. Broadway
New York, NY 10012
(212)925-2140

Ciba Consumer
Pharmaceuticals
581 Main Street
Woodbridge, NJ 07095
(908)602-6600

Citius USA, Inc.
120 Interstate North
Pkwy., East, Suite 106
Atlanta, GA 30339
(800)343-9099

Claylia's Own
440 Sawdust Rd.
The Woodlands, TX 77380

Clear Alternative
8707 West Ln.
Magnolia, TX 77355

Color Choice Division
701 Lake Ave.
Lake Worth, FL 33460

Colour Quest, Ltd.
1432 Lakeview Drive
Rochester Hills, MI
48306-4572

Columbia Manicure
Manufacturing Co.
1 Seneca Place
Greenwich, CT 06830

Combe, Inc.
1101 Westchester Avenue
White Plains, NY 10604
(800)431-2610

Compar, Inc.
Nine Skyline Drive
Hawthorne, NY 10532
(914)347-3680

ConAgra Pet Products
1 Century Park Plaza,
Suite 700
Omaha, NE 68102-1675
(402)595-7000

Connie Stevens/
Forever Spring
8721 Sunset Blvd.
Los Angeles, CA 90069

Contessa Cosmetics
2301 W. Sample Road,
Bldg. 3, Suite 6-A
Pompano Beach, FL 33073
(800)354-4247, (305)979-1298

Cosmetic & Fragrance
Concepts, Inc.
8839 Greenwood PL.
Savage, MD 20763

Cosmetic Chemist
(Naturel brand)
8108 Capwell
Oakland, CA 94621

Cosmetic Group
11340 Penrose St.
Sun Valley, CA 91352

Cosmetics Company
Trump Tower
725 Fifth Ave
New York, NY 10022-2519

Cosmetics Plus
1201 3rd Ave.
New York, NY 10021

Cosmo Pro, Inc.
320 Fentress
Daytona Beach, FL 32114
(904)254-1967

Creative Images, Inc.
6025 W. Monroe St.
Phoenix, AZ 85043

Crebel International
(TMV Corp)
4401 Ponce DeLeon Blvd.
Coral Gables, FL 33146
(800)457-4433

Critter Comfort
14200 Old Hanover Rd.
Reisterstown, MD 21136

Cumberland-Swan, Inc.
One Swan Dr.
Smyrna, TN 37167
(615)459-8900

CYA Products, Inc.
211 Robbins Lane
Syosset, NY 11791
(800)247-9274,
(718)937-2900

Dana Perfumes Corp.
635 Madison Ave.
New York, NY 10022
(212)751-3700

Dasco
P O Box 441
New York, NY 10150-0441

DeMert & Dougherty, Inc.
Five Westbrook Corp. Ctr.
Suite 900
Westchester, IL 60154
(800)532-5111

Dermac Mfg./Touch of Mink
P.O. Box 5268
Salem, OR 97304

Dermalogica, Inc.
1001 Knox Street
Torrance, CA 90502
(310)640-6882

Diamond Brands, Inc.
1660 S. Highway 100,
Suite 340
Minneapolis, MN 55416
(612)541-1500

DML Lotion
616 Allen Ave.
Glendale, CA 91201

Donna Karen New York
550 Seventh Avenue
New York, NY 10019
(212)789-1500

Dr. Moynahan
Skincare Center
18 Haynes St.
Hartford, CT 06103

Drs. Foster & Smith
2253 Air Park Road
Rhinelander, WI 54501-
0100
(800)323-4208

Duart Industries
1211 Flynn Road
Camarillo, CA 93012

DuCair Bioessence, Inc.
7777 Westside Ave.
North Bergen, NJ 07047

Duncan Enterprises
5673 East Shields Ave.
Fresno, CA 93727
(209)291-4444

Durham Pharmacal Corp.
Route 145
Oak Hill, NY 12460
(518)239-4195

Dynex International,
Ltd.
25 West 43rd St.
New York, NY 10036

E. Fougera & Co.
60 Baylis Road
Melville, NY 11747
(516)454-6996

Earth General
72 Seventh Avenue
Brooklyn, NY 11217
(800)562-2203

Earth Wise, Inc.
4600 Sleepytime Drive
Boulder, CO 80301-3292
(303)530-5300

echelle, Inc.
1236 Jordan Road
Huntsville, AL 35811

Echo-Logix
600 S. Curson Ave.
Suite 609
Los Angeles, CA 90036

Eco 1 Corporation
918 Sherwood Drive
Sherwood, IL 60044
(800)600-3261,
(708)831-2000

Eco Design Company
1365 Rufina Circle
Santa Fe, NM 87501
(800)621-2591,
(505)438-3448

Ecomer, Inc.
13 N. 7th St
Perkasie, PA 18944

Edward & Sons
Trading Company, Inc.
P.O. Box 1326
Carpinteria, CA 93014
(805)684-8500

Elas Fragrances, Inc.
999 E. 46th St.
Brooklyn, NY 10036

Eli Lilly & Company
Lilly Corporate Center
Indianapolis, IN 46285

Elizabeth Grady
Face First
200 Boston Ave., #3500
Medford, MA 02155-4243

Elkins Enterprises, Inc.
309 East Bailey St.
Globe, AZ 85501

Emlin Comsetics
290 Beeline Dr.
Bensenville, IL 60106-1611

enfasi Hair Care
2937 S. Alameda St.
Los Angeles, CA 90058
(800)326-3928,
(310)275-3928

Enforcer Products, Inc.

P.O. Box 1060
Cartersville, GA 30120
(800)241-5656

Equiflite Technologies, Inc.
1801 W. Rose Garden
Lane, Ste. 3
Phoenix, AZ 85027
(800)595-RACE

Erickson Cosmetic
Company
1920 N. Clybourn Ave.
Chicago, IL 60614

ES Laboratories
(Stevens Research Salon)
19009 61st Ave. NE, Unit 1
Arlington, VA 98223
(800)262,3344,
(206)435-4513

Esirg Manufacturing Co.
4901 Warehouse Way
Sacramento, CA 95826

Eubiotics, Ltd.
90 New York Ave.
West Hempstead, NY 11552

Excel-Mineral Co., Inc.
P.O. Box 3800
Santa Barbara, CA
93130-3800

Farnam Companies, Inc.
1302 Lew Ross Road
Council Bluffs, IA
51501-7703

Fashion & Designer
Fragrances, Inc.
625 Madison Ave.
New York, NY 10022

Fashion Fair Cosmetics
820 S. Michigan
Chicago, IL 60605

Feberge Organics
33 Benedict Place
Greenwich, CT 06830

Fleming Companies, Inc.
6301 Waterfront Blvd.,
P.O. Box 26647
Oklahoma City, OK 73126
(405)841-8308

Florida East Atlantic
Pet Products, Inc.
P.O. Box 8631
Coral Springs, FL 33075
(305)344-0131

Flowery Beauty Products
P O Box 4008
Greenwich, CT 06830
(203)661-0995

For Earth's Sake
11 Wyndham St. N.
North Guelph, Ontario,
Canada N1E 5Z7
(519)837-3242

For Women Only
1325 Eagandale Ct.
Eagan, MN 55121-1356

Forest Mist
Natural Products
9200 Mason Avenue
Chatsworth, CA 91311
(800)327-2012,
(818)882-2951

Forever Living Products
P.O. Box 29041
Phoenix, AZ 85038
(602)968-3999

Fragrance
Impressions, Ltd.
116 Knowlton St.
Bridgeport, CT 06608
(800)541-3204,
(203)367-6995

Frank Fuhrer International
3100 East Carson St.
Pittsburgh, PA 15203

Frasco Enterprises
One Fairchild Court
Plainview, NY 11803
(800)OMBRA 94

Frenchtop
Cosmetics USA
P.O. Box 92048
Elk Grove Village, IL
60009

Frontier
Box 299
Norway, IA 52318
(800)669-3275

Future Science Labs
P.O. Box 311
Mineola, NY 11501-0311
(800)257-2840

Galderma
3000 Alpha Mesa Blvd.
Suite 300
Ft. Worth, TX 76133
(817)263-2600

Gemma Italian
Herbal Skin Care
43 Grandview Drive
Trumbull, CT 06611

Gold Medal
Hair Products, Inc.
1 Bennington Ave.
Freeport, NY 11520

Golden Cat Corporation
300 Airport Road
P.O. Box 1086
Cape Girardeau, MO
63702-1086
(314)334-6618

Graham-Webb Int'l
13700 1st Avenue
Plymouth, MN 55441
(800)456-9322

Greentree
Laboratories, Inc.
P.O. Box 425
Tustin, CA 92681
(415)546-9520

Guardsman Products, Inc.
P.O. Box 1521
Grand Rapids, MI 49501
(616)957-2600

Gucci Parfums
15 Executive Blvd.
Orange, CT 06477

Guthy-Renker Corporation
7 Holland
Irvine, CA 72718-2506
(800)545-5595

H.A. Cole
Products Company
P O Box 9937
Jackson, MS 39286
(800)235-PINE

Happy Jack, Inc.
Box 475
Snow Hill, NC 28580
(800)326-5226

Hartz Moutain
Corporation
700 Frank E. Rodgers
Blvd., S.
Harrison, NJ 07029

Health & Body Fitness, Inc.
12021 Wilshire Blvd.,
Suite 621
West Los Angeles, CA
90025
(310)434-6417

Heartland Scent
7782 Newburg Road
Newburg, PA 17240-9601
(800)323-0418

Hermes of Paris, Inc.
745 Fifth Ave., #800
New York, NY 10151-0123
(212)759-7585

Hewitt Soap Co., Inc.
333 Linden Ave.
Dayton, OH 45403
(800)543-2245

Hill Dermaceuticals, Inc.
P.O. Box 149283
Orlando, FL 32814
(407)896-8280

Hillestad
International, Inc.
P.O. Box 41218
San Jose, CA 95160-1218
(408)298-0995

Holistic Products Corp.
10 W. Forest Ave.
Englewood, NJ 07631
(800)221-0308

Imperial-Dax
P O Box 10002
Fairfield, NJ 07004
(201)227-6105

Incodisa Soap
and Cosmetics
56 Summitt Avenue
Tappan, NY 10983

Inne Dispensables, Inc.
144 Fairfield Road
Fairfield, NJ 07004-2406
(516)273-5100

International Oriental
Beauty Secrets
1800 S. Robertson Blvd.,
#182
Los Angeles, CA 90035

Inverness Corporation
17-10 Willow Street
Fair Lawn, NJ 07410
(201)794-3400

Jack Eckerd Corporation
8333 Bryan Dairy Road
Largo, FL 34647
(813)397-7461

Jackie Brown Cosmetics
2122 Anthony Drive
Tyler, TX 75701
(903)561-5053

Jaclyn Cares
P.O. Box 514, Dept. M
Farmington, MI 48332-0514
(313)453-5410

James River Co.
120 Tredegar St.
Richmond, VA 23217
(203)854-2000

Jerome Milton, Inc.
4350 W. Ohio St.
Chicago, IL 60624
(312)638-1800

Jeunesse Natural
Skin Care
22525 Cass Ave.
Woodland Hills, CA
92911

Jewel Company
1955 W. North Avenue
Melrose Park, IL 60160
(708)531-6000

JK Templeton, Inc.
598 Display Way
Sacramento, CA 95834
(916)920-4788

John F. Amico & Co.
7327 W. 90th St.
Bridgeview, IL 60455

Johnson Products Co., Inc.
8522 S. Lafayette Ave.
Chicago, IL 60620
(800)621-6043

Jones
21 Seldom Seen Road
Bradfordwoods, PA
15015-1321

Jose Eber, Inc.
888 7th Ave., 43rd Fl.
New York, NY 10106

Karina, Inc.
12 Van Vooren Dr.
Oakland, NJ 07436

Kash'N'Karry
6422 Harney Road
Tampa, FL 33610
(800)882-2505

KB Products, Inc.
20 N. Railroad Ave.
San Mateo, CA 94401
(800)464-5458,
(415)344-6500

Kemi Labs
9520 Gerwig Lane
Columbia, MD 21046
(410)381-6664

Ken Lange
7112 N. 15th Place, Suite 1
Phoenix, AZ 85020
(800)486-3033

Kitty Express
144-A South Valencia St.
Glendora, CA 91741
(800)848-1323

Knomark, Inc.
328 Gale St.
Aurora, IL 60506

L & F Household Products
225 Summit Ave.
Montvale, NJ 07645
(800)888-0192

La Prairie, Inc.
31 West 52nd St.
New York, NY 10019-6118
(212)459-1600

Lambert-Kay Division
Half Acre Road
Cranbury, NJ 08512
(609)426-4800

Lancaster Group USA
745 Fifth Avenue
New York, NY 10151
(212)593-7400

Lander Company, Inc.
106 Grand Avenue
Englewood, NJ 07631
(201)568-9700

Le Salon Paul Morey
1301 5th Avenue, Suite 10
Seattle, WA 98101-2603

Leiner Health Products
901 E. 233rd Street
Carson, CA 90745-6204

Lewis Laboratories
Int'l., Ltd.
49 Richmondville Avenue
Westport, CT 06880

Liberty Natural
Products, Inc.
8120 S.E. Stark Street
Portland, OR 97216
(800)289-8427

Light Force
115 Thompson Ave., #5
Santa Cruz, CA 95062
(408)464-6500

Lightning Products
10100 NW Executive
Hills Blvd.,#105
Kansas City, MO 64153
(800)422-8858

Lissee Cosmetics
2937 S. Alameda Street
Los Angeles, CA 90058
(800)326-3974, (310)277-3974

Loctite Corporation
4450 Cranwood Parkway
Cleveland, OH 44128-4084
(216)475-3600

Luster Products Co., Inc.
1631 S. Michigan Ave.
Chicago, IL 60616

Lynae/Boscogen, Inc.
11 Morgan Street
Irvine, CA 92718
(714)380-4317

Marae Storm
P.O. Box 203
Branscomb, CA 95417
(707)984-8755

Markwins International
Corporation
1931 Yeager Avenue
LaVerne, CA 91750
(909)392-0818

Martin F. Weber Company
2727 Southhampton Road
Philadelphia, PA 19154
(215)677-5600

Matol Botanicals
International
1111 46th Avenue
Montreal, Quebec,
Canada H8T 3C5

Maybelline, Inc.
P.O. Box 372
Memphis, TN 38101-0372
(901)324-0310

MBL Industries, Inc.
990 Industrial Park Drive
Marietta, GA 30062

McKesson Corporation
One Post St.
San Francisco, CA
94104
(800)747-4104

McNeil Consumer
Products Co.
Div. Of McNeil-PPC Inc.
Ft. Washington, PA 19034
(215)233-7000

MDR Fitness Corporation
14101 NW 4th Street
Fort Lauderdale, FL
33325-6209

Medtech Laboratories, Inc.
P.O. Box 1108
Jackson, WY 83001
(800)443-4908

Mehron
100 Red School House Rd.
Spring Valley, NY 10977
(914)426-1700

Michelle Lazar Cosmetics
755 S. Lugo Avenue
San Bernardino, CA 92048
(800)676-8008,
(909)888-6310

Mira Linder
Spa in the City
29935 Northwestern Hwy.
Southfield, MI 48034
(800)321-8860, (313)356-5810

Mon Ami Industrial Co.
P.O. Box 5217
Los Angeles, CA 90055-0217

Mothercare
4255 N. Knox Avenue
Chicago, IL 60641-1904

Nailtiques
2897 SE Monroe Street
Stuart, FL 34997-5914
(800)272-0034, (305)378-0740

Nala Berry Labs
P.O. Box 151
Palm Desert, CA 92261
(800)397-4174, (619)568-6970

National Laboratories
Chestnut Ridge Plaza
Montvale, NJ 07645
(201)573-5280

Natural Farms
429 H St.
Los Banos, CA 93635

Natural Organics, Inc.
548 Broadhollow Road
Melville, NY 11747
(800)645-9500, (516)293-0030

Nature's Apothacary
6530 Gunpark Dr., Ste 500
Boulder, CO 80301
(800)999-7422

Nature's Bounty, Inc.
90 Orville Dr.
Bohemia, NY 11716
(800)645-5412

Nature's Design, Ltd.
22 Woodhull St.
Brooklyn, NY 11231
(718)522-4250

Nature's Way Products, Inc.
10 Mountain Springs Pkwy.
Springville, UT 84663
(801)489-1500

Naturel (Surrey)
13110 Trails End Road
Leander, TX 78641
(512)267-7172

Neocare Laboratories, Inc.
3333 W. Pacific Coast Hwy.
Newport Beach, CA 92663
(800)982-4NEO

Neoteric Cosmetics
4880 Havana Street
Denver, CO 80239
(800)55-ALPHA,
(303)373-4860

Neutrogena Skincare Institute
5760 West 96th St.
Los Angeles, CA 90045
(800)421-6857

Nice Pack Products, Inc.
2 Nice Pack Park
Orangeburg, NY 10962

Nordstrom Cosmetics
P1321 2nd Avenue
Seattle, WA 98111-4338
(800)7-BEAUTY,
(206)233-5569

NOVATREND
2810 Temple Avenue
Long Beach, CA 90806
(310)426-1705

NOW Products
1045 S. Edward Drive
Tempe, AZ 85281
(800)662-0333

NR Laboratories, Inc.
900 E. Franklin Street
Centerville, OH 45459
(800)223-9348,
(513)433-9570

Nu Skin Int'l., Inc.
75 West Center Street
Provo, UT 84601
(800)877-6100

Nu-Nails
7241 W. Lunt Avenue
Chicago, IL 60631-1122

Oil of Orchid
P.O. Box 1040
Guernerville, CA 95446
(707)869-0761

One Earth Products
P.O. Box 1331
Santa Barbara, CA 93102

Only Natural, Inc.
14 Buchman Road
Salem, MA 01970
(508)745-9766

Optkin Int'l
P.O. Box 27319
Denver, CO 80227

Oriflame Corporation
76 Trebel Cove Road
North Bellerica, MA 01862
(508)663-2700

Osco Drugs
1818 Swift Drive
Oakbrook, IL 60521
(708)572-5000

Otto Basics Beauty 2 Go!
P.O. Box 9023
Rancho Santa Fe, CA 92067
(800)598-OTTO,
(619)756-2026

Owen/Gaiderma
Labs/ Cetaphil
P O Box 6600
Fort Worth,TX 76115

Pacific Gold Coast Co.
1455 Holiday Hill Road
Goleta, CA 93117-1836
(805)683-4233

Paco Rabanne Parfums
70 E. 55th St.
New York, NY 10022

Pamela Marsen, Inc.
P O Box 119
Teaneck, NJ 07666
(201)836-7820

Pantresse, Inc.
65 Schawmut Road
Canton, MA 02021
(617)828-7199

Parfums De Coeur, Ltd.
85 Old Kings Highway, N.
Darien, CT 06820
(203)655-8807

Park Rand Enterprises, Inc.
13143 1/2 San Fernando Rd.
Sylmar, CA 91342-3542
(818)362-6218

Parke-Davis
2800 Plymouth Rd
Ann Arbor, MI 48105
(313)996-7324

Pathmark (Supermarkets
General Corp.)
301 Blair Road
Woodbridge, NJ 07095

Pavlon, Ltd.
Nyack-on-the-Hudson,
NY 10960

Perfumers Workshop, Ltd.
18 E. 48th Street
New York, NY 10017
(212)759-9491

Perrigo Company
117 Water
Allegan, MI 49010
(616)673-8451

Personal Products Co.
199 Grandview Road
Skillman, NJ 08558
(800)526-3967

Pet Organics/Nala
Laboratories
P.O. Box 151
Palm Desert, CA 92261-0151

Pet-E-Coat USA, Inc.
210 Terminal Drive
Plainveiw, NY 11803
(516)349-0780

Pfizer, Inc.
235 East 42nd Street
New York, NY 10017
(212)573-2323

Physician's Formula
Cosmetics, Inc.
230 South 9th Ave.
City of Industry, CA 91746
(800)227-0333

Pilkington Barnes
Hind, USA
1100 East Bell Rd
Phoenix, Az 85022-2696
(800)528-7971

Planet Products, Inc.
26 Silkberry
Irvine, CA 92714-7480
(310)417-8468,
(604)656-9436

Players Edge
P.O. Box 429
Hickory, NC 28603

Price Costco
10809 120th Avenue NE
Kirkland, WA 98033
(800)776-6702,
(206)828-8100

Principal Secret
5950 La Place Court, Ste. 160
Carlsbad, CA 92008
(800)748-5780

Pro-Line
2121 Panaramoric Circle
Dallas, TX 75212
(214)631-4247

Professional Choice
Hair Care
2937 S. Alameda Street
Los Angeles, CA 90058
(800)326-3974,
(310)277-3974

PSI Industries, Inc.
P.O. Box 4391
Roanoke, VA 24015
(703)345-5013

Pulse Products
2021 Ocean Ave., #105
Santa Monica, CA 90405

Pure Touch
Therapeutic Body Care
P.O. Box 1281
Nevada City, CA 95959
(800)442-PURE,
(916)265-5949

Quantum, Inc.
754 Washington Street
Eugene, OR 97401
(800)448-1448

Quintessence, Inc.
980 N. Michigan Ave.
Chicago, IL 60611
(312)951-7000

Rathdowney, Ltd.
3 River Street
Bethel, VT 05032
(800)543-8885

Red Rose Collection, Inc.
42 Adrian Ct.
Burlingame, CA 94010

Renaissance Skin Care, Inc.
1456 Second Ave., Ste. 105
New York, NY 10021

Richard Ravid, Inc.
4250 Normandy Court
Royal Oak, MI 48073
(313)549-3350

Rite-Aid Corporation
30 Hunter Lane
Camphill, PA 17011
(717)761-2633

Riviera Concepts, Inc.
150 Duncan Mill Road
Don Mills, Ontario,
Canada M3B 3M4
(416)441-9933

Robert Research
1480 SW 30th Avenue
Boynton Beach, FL 33426
(407)734-4062

Rochas Paris, Inc.
29 West 57th St.
New York, NY 10019

Roux Labs
5344 Overmyer Drive
Jacksonville, FL 32205
(904)693-1200

Royal Laboratories
2849 Dundee Road,
Suite 112
Northbrook, IL 60062
(800)876-9253

Rusk
606 West Mount Dr.
Los Angeles, CA 90069

Saabune Products Inc.
560 Allyn Street
Akron, OH 44311
(216)535-8543

Safetex Corporation
2926 Columbia Hwy.
Dothan, AL 36303-5406
(800)520-8341

Safeway
4th and Jackson Street
Oakland, CA 94660

Salis International, Inc.
4093 N. 28th Way
Hollywood, FL 33020
(305)921-6971

Sani-Wax, Inc.
8340 Mission Rd., Ste. 102
Prairie Village, KS
66206-1362

Scandinavian Natural
Health & Beauty
Products, Inc.,
13 N. 7th Street
Perkasie, PA 18944
(800)288-2844

Scentura Creations
5616 Peachtree Rd.
Atlanta, GA 30341

Scott's Liquid Gold
4880 Havana St.
Denver, CO 80239
(303)373-4860

Shark Products
Brooklyn Navy Yard,
Flushing & Cumberland St.
Brooklyn, NY 11205
(718)624-6161

Shirlo, Inc.
4242 BF Goodrich Blvd.
Memphis, TN 38118
(901)362-1950

Signature Beauty Care, Inc.
1111 SW 30th Ave., 2nd Flr.
Deerfield Beach, FL
33442-8154
(305)864-1044

Simmons Soaps
42295 E. Highway 36
Bridgewille, CA 95526

Simple Green
15922 Pacific Coast Hwy.
Huntington Harbor, CA
92649
(800)228-0709

Siri Skin Care, Inc.
4305 1/2 Eagle Rock Blvd.
Eagle Rock, CA 90041
(213)254-5628

Smith Enterprises, Inc.
1953 Langston Street
Rock Hill, SC 29731
(803)366-7101

Smith's Stores
1550 S. Redwood Road
Salt Lake City, UT 84104
(801)974-1400

Soft Sheen
Products, Inc.
1000 East 87th St.
Chicago, IL 60619
(312)978-0700

Sorik International
278 Taileyand Avenue
Jacksonville, FL 32202
(800)940-HAIR,
(904)353-4200

Soya Systems, Inc.
1572 Page Industrial Dr.
St. Louis, MO 63132
(314)428-0004

Spectrum Group
8494 Champin Industrial Dr.
St. Louis, MO 63114
(314)427-4886

St. Ives Laboratories, Inc.
9201 Oakdale Avenue
Chatsworth,CA 91311
(800)421-9231

St. Jon Laboratories
1656 W. 240th Street
Harbor City, CA 90710
(800)969-7387

Stacy Allan Cosmetics, Inc.
P.O. Box 13034
Milwaukee, WI 53213
(414)744-4377

Stanhome, Inc.
333 Western Avenue
Westfield, MA 01085
(413)562-3631

Stetson
P.O. Box 241
Church Hill, MD 21690

Stevens Research
19009 61st Ave., NE,
Dept. 1
Arlington, WA 98223

Stop'N'Shop Products
P.O. Box 1942
Boston, MA 02105

Straight Arrow Products
9201 Oakdale Avenue
Chatsworth, CA 91311

Super G, Inc.
6300 Sheriff Road
Landover, MD 20785
(301)341-4102

Supreme Beauty Products
820 South Michigan
Chicago, IL 60605

Surrey
13110 Trails End Road
Leander, TX 78641
(512)267-7172

Tanning Research
Laboratories
P.O. Box 5111
Daytona Beach, FL 32118
(904)677-9559

Target Stores
33 South Sixth Street
Minneapolis, MN
55440-1392
(612)370-6000

Tranzonic Companies
P.O. Box 94763
Cleveland, OH 44101-4763

Trophy Animal Health Care
2796 Hlen Street
Pensacola, FL 32504
(800)336-7087

Tropical Botanicals, Inc.
P.O. Box 1354
Rancho Santa Fe, CA 92067

Tsumura International, Inc.
300 Lighting Way
Secaucus, NJ 07096
(201)223-9000

Twin Laboratories, Inc.
2120 Smithtown Avenue
Ronkonkoma, NY 11779
(516)467-3140

Tyra Skin Care, Inc.
9019 Oso Avenue, Ste. A
Chatsworth, CA 9131
(800)322-TYRA,
(818)407-1274

UAS Laboratories
5610 Rowland Road, Ste. 110
Minnetonka, MN 55343
(612)935-1707

Ultima Brands USA, Inc.
1913 Atlantic Ave.
Manasquan, NJ 08736-1094

Ultimate Life
P.O. Box 31154
Santa Barbara, CA 93130
(800)594-9333,
(805)962-7556

Universal Labs
3 Terminal Road
New Brunswick, NJ 08901
(800)872-0101,
(908)545-3130

Vanilla Fields Body Care
P.O. Box 5574
Newton, CT 06470-5574

Velvet Products Company
P.O. Box 5459
Beverly hills, CA 90210
(213)472-6431

Veterinarian's Best
1554 Knoll Circle Dr.
Santa Barbara, CA 93103

Vickie Lavanty International
5697 LaJolla Blvd., Suite 1B
LaJolla, CA 92037
(619)551-8700

Victoria Vouge
90 Southland Drive
Bethleum, PA 18017
(800)967-PUFF

Vienna Beauty Products
P.O. Box 211, North
Dayton Station
Dayton, OH 45404
(513)228-7109

Vita-Tech International, Inc.
2832 Dow Avenue
Tustin, CA 92680
(714)832-9700

Vitae
2404 Stayton Way
Louisville, KY 40202

Wakefern Food
Corporation (Shop-Rite)
600 York Street
Elizabeth, NJ 07207

Wakunaga of America
Company, Ltd.
23501 Madero
Mission Viejo, CA 92691
(800)421-2998

Warner-Lambert Company
201 Tabor Rd.
Morris Plains, NJ 07950

Watkins, Inc.
P.O. Box 5570
Winona, MN 55987
(507)457-3300

Whitehall Labs
5 Giralda Farms
Madison, NJ 07940
(201)660-5500

Withers Mill Company
P.O. Box 347
Hannibal, MO 63401
(800)223-0858

Woolworth, FW
233 Broadway
New York, NY 10279
(212)553-2000

Wyeth Laboratories, Inc.
P.O. Box 8299
Philadelphia, PA 19101
(800)666-7248

Yardley of London, Inc.
3030 Jackson Ave.
Memphis, TN 38112-2018

❤ PRODUCT REFERENCE GUIDE ❤

This section will assist you in finding specific products manufactured by the **cruelty-free companies** listed in this book. Keep in mind that not every manufacturer and/or distributor is listed in this section.

Air Fresheners
Amberwood
American Eco-Systems
Aromaland, Inc.
Body & Soul of Chicago
Breezy Balms
Caswell-Massey
Cinema Secrets, Inc.
Dial Corp.
Ecco Bella Botanicals
EM Enterprises
Gryphon Development
Hair Doc Company
Herbal Products &
    Development
Heritage Store
Infinite Quality, Inc.
IQ Products Company
Les Femmes, Inc.
Leydet Aromatics
Lotions & Potions
McLaughlin, Inc.
Medicine Flower, Inc.
Mother's Little Miracle, Inc.
Mountain Rose Herbs
Nectar USA
New Methods
Perfect Natural Products
Plant Essence
Precious Collection
Sarah Michaels
Seventh Generation
Surco Products, Inc.
TerrEssentials
Thursday Plantation, Inc.
Vegan Market
Vermont Country Soap
Zenith Advanced Health
    Systems Int'l, Inc.

Aromatherapy
Alexandra Avery Purely
    Natural Body Care

Aloe Creme Laboratories
American International Ind.
Ananda Country Products
Aphrodisia Products, Inc.
Aroma Life Company
Aromaland, Inc.
Avalon Natural
    Cosmetics, Inc.
Aveda Corporation
Bath Island, Inc.
Bavarian Alpenol &
    Sunspirit
BeautiControl
    Cosmetics, Inc.
Bella's Secret Garden
Bio-Tec Cosmetics, Ltd.
Black Pearl Gardens
Blessed Herbs
Body & Soul of Chicago
Body Encounters
Body Love Natural
    Cosmetics, Inc.
Body Shop, Inc.
Body Time
Body Tools
Bodyography
Botan Corporation
Brookside Soap Company
Canada's All
    Natural Soap, Inc.
Chishti Company
Christopher Enterprises
Classic Cosmetics, Inc.
Cleopatra's Secret
    from the Dead Sea
Colin Ingram
Common Scents
Compassion Matters
Creation Soaps, Inc.
Decleor USA, Inc.
Desert Essence
Devi, Inc.

Earth Science, Inc.
Ecco Bella Botanicals
Eden Botanicals
EM Enterprises
Essential Aromatics
Essential Elements
Essential Products
    of America, Inc.
Exotic Nature
Faith Products, Ltd.
Forest Essentials
Garcoa Labs
Gena Labs
Helen Lee Skin
    Care & Cos.
Heritage Store, Inc.
Internatural
Izy's Skin Care Products
Janca's Jojoba Oil &
    Seed Company
Katonah Scentral
Khepra Skin Care, Inc.
Kiss My Face Corp.
L'Herbier De Provence, Ltd.
La Dove, Inc.
La Natura
Lady of the Lake Company
Leydet Aromatics
Little Herbal Garden
Lotions & Potions
Lotus Brands
Lotus Light Enterprises
Lunar Farms Herbal
    Specialist
Luzier Personalized
    Cosmetics
Maharishi Ayur-Veda
    Products
Mar-Riche Enterprises, Inc.
Marilyn Miglin, L.P.
Masada Marketing Company
Master's Flower Essences
Medicine Flower, Inc.

Mera Personal
Care Products
Montagne Jeunesse
Mountain Rose Herbs
Nadina's Cremes
Native Scents, Inc.
Natural Attitudes, Inc.
Natural Bodycare
Natural Way Natural
Body Care
Nature's Corner
Nature's Elements
International, Ltd.
NaturElle Cosmetics Corp.
Nectar USA
New Moon Extracts, Inc.
Neways, Inc.
Paul Mazzotta, Inc.
Pepper's
Phybiosis
PlantEssence
Potions & Lotions/
Body & Soul
Precious Collection
Prestige Cosmetics
Quan Yin Essentials
Ravenwood
Reviva Labs, Inc.
RJF, Inc./Oshadhi
Royal Labs Natural
Cosmetics
Set-N-Me-Free
Seventh Generation
Shirley Price
Aromatherapy Ltd.
Simplers Botanical Co.
Sinclair & Valentine
Smith & Vandiver, Inc.
Starwest Botanicals, Inc.
Strong Skin Savvy, Inc.
Sumeru Garden Herbals
Sunfeather Herbal
Soap Company
Surco Products, Inc.
Thursday Plantation
Labs, Pty., Ltd.
Tisserand Aromatherapy
Vegan Market
Venus & Apollo

Vermont Country Soap
Virginia's Soap Limited
Whole Spectrum
Windrose Trading Company
WiseWays Herbals

Baby Products
Action Labs
Alexandra Avery Purely
Natural Body Care
Alivio Products, Inc.
Aloe Creme Laboratories
Amberwood
Aubrey Organics
Autumn Harp, Inc.
Aware Diaper, Inc.
Bare Escentuals
Basically Natural
Bath Island, Inc.
Baudelaire, Inc.
Bella's Secret Garden
Blessed Herbs
Body & Soul of Chicago
Borlind of Germany
Botan Corporation
Brookside Soap Company
C.E. Jamieson & Co., Ltd.
Canada's All Natural
Soap, Inc.
Caring Catalog
Carme, Inc.
Chanty, Inc.
Cinema Secrets, Inc.
Classic Cosmetics, Inc.
Compassion Matters
Country Comfort
Earth Science, Inc.
Faith Products, Ltd.
Garcoa Labs
Garden Botanika
Good Clean Fun
Granny's International, Inc.
Great Mother's Goods
Green Mountain
Products, Inc.
Healthy Times
Infinite Quality, Inc.
Internatural
Katonah Scentral

La Dove, Inc.
La Natura
Logona USA, Inc.
Lotus Brands
Lunar Farms Herbal
Specialist
Magick Botanicals/
Magick Mud
Mastey De Paris, Inc.
McLaughlin, Inc.
Mother Love
Herbal Company
Mother's Little
Miracle, Inc.
Mountain Rose Herbs
Naturade
Natural Bodycare
Nature's Corner
Nature's Elements
International, Ltd.
Nectar USA
North Country Glycerin
Soap and Candle
Palm Beach Beauty Products
Perfect Natural Products
Royal Labs Natural
Cosmetics
Seventh Generation
Shaklee U.S., Inc.
Strong Skin Savvy, Inc.
Sumeru Garden Herbals
Sunfeather Herbal
Soap Company
Sunrise Lane Products, Inc.
Terra Flora Herbal Body
Care Products
TerraNova
Terressentials
Thursday Plantation
Labs, Pty., Ltd.
Venus & Apollo
WiseWays Herbals

Bath Products
Action Labs
AKA Saunders, Inc.
Alba Naturals, Inc.
Alexandra Avery Purely
Natural Body Care

Alivio Products, Inc.
All-One-God-Faith, Inc.
Aloe Flex Products, Inc.
Amberwood
American ORSA, Inc.
Ananda Country
    Products
Andrew Jergens Company
Aphrodisia Products, Inc.
Aromaland, Inc.
Aubrey Organics
Auromere Ayurvedic
    Imports
Avalon Natural
    Cosmetics, Inc.
Aveda Corporation
Bare Escentuals
Basically Natural
Bath Island, Inc.
Baudelaire, Inc.
BeautiControl
    Cosmetics, Inc.
Bella's Secret Garden
Benetton Cosmetics Corp.
Bio-Pac
Bio-Tec Cosmetics, Ltd.
Biogime
Black Pearl Gardens
Body/Body, Bath &
    Body Products
Body & Soul of Chicago
Body Shop, Inc.
Body Time
Body Tools
Bodyography
Bonne Bell, Inc.
Borlind of Germany
Botan Corporation
Bradford Soap Works, Inc.
Brookside Soap Company
Canada's All Natural
    Soap, Inc.
Caring Catalog
Carme, Inc.
Caswell-Massey
Chanty, Inc.
Chempoint Products, Inc.
CHIP Distribution Company
Cinema Secrets, Inc.

Clarins of Paris
Classic Cosmetics, Inc.
Clearly Natural Products
Cleopatra's Secret
    from the Dead Sea
Cloudworks
Coastline Products
Colonial Dames Co. Ltd.
Color My Image, Inc.
Compassion Matters
Conair Corporation
Concept Now Cosmetics
Creation Soaps, Inc.
Creme de la Terre
Delby System
Dena Corporation
Dep Corporation
Desert Essence
Devi, Inc.
Dial Corporation
Dr. Hauschka Cosmetics
E. Burnham Cosmetic
    Company, Inc.
Earth Preserv, Ltd.
Earth Science, Inc.
Ecco Bella Botanicals
Ella Bache, Inc.
EM Enterprises
Essential Aromatics
Essential Elements
Essential Products
    of America, Inc.
Essque Bodycare
European Soaps, Ltd.
Eva Jon Cosmetics
Ever Young, Inc.
Exotic Nature
Faith Products, Ltd.
Finelle Cosmetics
Forest Essentials
French Transit, Ltd.
Garcoa Labs
Garden Botanika
Gena Labs
Goodier, Inc.
Granny's International, Inc.
Great Mother's Goods
Green Mountain
    Products, Inc.

Greenway Products, Inc.
Gryphon Development
Hair Doc Company
Hawaiian Resources Co., Ltd.
Head Shampoo, Inc./
    Pure & Basic Products
Healthy Times
Helen Lee Skin Care & Cos.
Herb Garden
Heritage Storage
Homebody/Perfumoils, Inc.
Internatural
Izy's Skin Care Products
Janca's Jojoba Oil &
    Seed Company
JR Liggett, Ltd.
Kama Sutra Company
Katonah Scentral
Kiss My Face Corp.
KMS Research, Inc.
KSA Jojoba
L'Herbier De Provence, Ltd.
La Dove, Inc.
La Natura
Lady of the Lake Company
Levlad, Inc.
Leydet Aromatics
Little Herbal Garden
Logona USA, Inc.
Lotions & Potions
Lotus Brands
Lotus Light Enterprises
Louise Bianco Skin
    Care, Inc.
Lunar Farms Herbal
    Specialist
Luzier Personalized
    Cosmetics
Magic of Aloe, Inc.
Mar-Riche Enterprises, Inc.
Masada Marketing Company
Mastey De Paris, Inc.
McLAughlin, Inc.
Medicine Flower, Inc.
Michel Constantini Cosme
Montagne Jeunesse
Mother Love
    Herbal Company
Mountain Rose Herbs

Murad, Inc.
Nadina's Cremes
Native Scents, Inc.
Naturade
Natural Attitudes, Inc.
Natural Bodycare
Nature's Acres
Nature's Corner
Nature's Elements
    International, Ltd.
NaturElle Cosmetics Corp.
Nectar USA
Neways, Inc.
Norimoor Company, Inc.
North Country Glycerin
    Soap and Candle
Oasis Brand Products
Palm Beach Beauty Products
Patricia Allison
Paul Mazzotta, Inc.
Penn Herb Company, Ltd.
Pepper's
Phybiosis
Potions & Lotions/
    Body & Soul
Precious Collection
Ravenwood
Reviva Labs, Inc.
RJF, Inc./Oshadhi
RJG, Inc. (Formerly
    Dymer & Associates)
Royal Labs Natural
    Cosmetics
San Francisco
    Soap Company
Sappo Hill Soapworks
Sarah Michaels
Schroeder & Tremayne Inc.
Sea Minerals Co.
Set-N-Me-Free
Seventh Generation
Shaklee U.S., Inc.
Sierra Dawn Products
Simple Wisdom, Inc.
Sinclair & Valentine
Smith & Vandiver, Inc.
Soap Factory
Soap Works
Starwest Botanicals, Inc.

Steps In Health, Ltd.
Sumeru Garden Herbals
Sunfeather Herbal
    Soap Company
Sunrise Lane Products, Inc.
TCCD International, Inc.
Terra Flora Herbal
    Body Care Products
Terressentials
Thursday Plantation
    Labs, Pty., Ltd.
Tisserand Aromatherapy
Tom's of Maine
Travel Mates America
Tropical Soap Company
U.S. Sales Service
Uncommon Scents, Inc.
Vanda Beauty Counselor
Vegan Market
Vermont Country Soap
Virginia's Soap Limited
WiseWays Herbals

**Bathroom/Kitchen
Cleaners**
AFM Enterprises
Allens Naturally
Aloeprime, Inc.
Amazon Premium Products
Amberwood
America's Finest
    Products Corp.
American Eco-Systems
Austin's
Aveda Corporation
Basically Natural
Beaumont Products, Inc.
Bio-Pac
Biogime
Caring Catalog
Chempoint Products, Inc.
Citra Brands, Inc.
Coastline Products
Dial Corporation
Earth Friendly Products
Earthly Matters
Ecover, Inc.
EM Enterprises
Faith Products, Ltd.

Faultless Starch/Bon
    Ami Company
Golden Pride - Rawleigh
Green Mountain
    Products, Inc.
Greenway Products, Inc.
Home Service
    Products Company
Huish Detergents, Inc.
Jelmar
Lotus Brands
McLaughlin, Inc.
Neways, Inc.
Oasis Biocompatible
Pro-Ma Systems (USA), Inc.
RCN Products, Inc.
Seventh Generation
Shaklee U.S., Inc.
Sierra Dawn Products
Simple Wisdom, Inc.
Sunfeather Herbal
    Soap Company
Sunrise Lane Products, Inc.
Third Millennium Science
Unelko Corporation
Vegan Market
Zenith Advanced Health
    Systems Int'l, Inc.

Carpet/Rug Care
AFM Enterprises
Amberwood
American Eco-Systems
Amway Corporation
Austin's
Bio-Pac
Card Corporation
Caring Catalog
Chempoint Products, Inc.
CHIP Distribution Company
Coastline Products
Color & Herbal Company
Dial Corporation
Earthly Matters
EM Enterprises
Frank T. Ross & Sons, Ltd.
Granny's International, Inc.
Green Mountain Products
Huish Detergents, Inc.

Lan-O-Sheen, Inc.
McLaughlin, Inc.
New Methods
Seventh Generation
Simple Wisdom, Inc.
Soap Factory
Val-Chem Company, Inc.
Vegan Market

Companion Animal Care
Alpen Limited
Amberwood
American Eco-Systems
Ampro Industries
Ancient Formulas, Inc.
Aubrey Organics
Basically Natural
Blue Ribbons Pet Care
Brookside Soap Company
C.E. Jamieson & Co., Ltd.
Caring Catalog
Color & Herbal Company
Creation Soaps, Inc.
Dr. AC Daniels
Dr. Goodpet
Eco Pak Canada
Ellon USA
EM Enterprises
Espree Associates, Inc.
Eva Jon Cosmetics
Four (IV) Trail Products
Green Ban
Greenway Products, Inc.
Hansen's Pet
    Products Company
Herb Garden
Horseman's Dream
Janca's Jojoba Oil &
    Seed Company
KSA Jojoba
Mallory Pet Supplies
Master's Flower Essences
McLaughlin, Inc.
MediPatch Laboratories
    Corp.
Morrill's New Direction
Mountain Rose Herbs
Natural Animal, Inc.

Natural Research
    People, Inc.
Naturally Yours, Alex
Nature's Country Pet
Nature's Plus
New Methods
North Country Glycerin
    Soap and Candle
Oil-Dri Corp. of America
One World Botanicals
Oxyfresh USA, Inc.
Palm Beach Beauty
    Products
Pepper's
Pet Connection
Pet Guard
Pet Tech
Pets 'N' People, Inc.
Pro-Tec Pet Health
Professional Pet Products
Real Animal Friends
Rio Vista Equine Products
Rx for Fleas
Safer Chemical Co.
Set-N-Me-Free
Simplers Botanical Co.
Soap Works
Sojourner Farms Natural
    Pet Products
Solid Gold Holistic Animal
    Equine Nutrition Ctr.
St. Clair Industries, Inc.
Sunfeather Herbal
    Soap Company
Super Dry Industries
Terressentials
Third Millennium Science
Thursday Plantation
    Labs, Pty., Ltd.
Vegan Market
Wysong Corporation

Condoms
Body Tools
Finley Pharmaceuticals
Sagami, Inc.

Contact Lens Care
Lobob Laboratories, Inc.

Cosmetics
Alivio Products, Inc.
Aloe Creme Laboratories
Aloe Flex Products, Inc.
Aloeprime, Inc.
Amberwood
American International Ind.
Amway Corporation
Aroma Life Company
Aubrey Organics
Avalon Natural
    Cosmetics, Inc.
Aveda Corporation
Avon Products, Inc.
Bare Escentuals
Bath Island, Inc.
BeautiControl
    Cosmetics, Inc.
Beauty Without
    Cruelty Cosmetics
BeautyMasters
Benetton Cosmetics Corp.
Blue Cross Beauty
    Products, Inc.
Bob Kelly Cosmetics
Body & Soul of Chicago
Body Shop, Inc.
Body Time
Bodyography
Bonne Bell, Inc.
Borlind of Germany
Botan Corporation
Botanics of California
Botanics Skin Care
C.E. Jamieson & Co., Ltd.
Caring Catalog
Carma Laboratories, Inc.
Carme, Inc.
Caswell-Massey
Chanty, Inc.
Chenti Products, Inc.
Christian Dior Perfumes, Inc.
CiCi Cosmetics
Cinema Secrets, Inc.
Clarins of Paris
Classic Cosmetics, Inc.
Cleopatra's Secret
    from the Dead Sea
Clientele, Inc.

Colonial Dames Co., Ltd.
Color Me Beautiful
Color My Image, Inc.
Colora
Comfort Mfg. Company
Compassion Matters
Concept Now Cosmetics
Delby System
Dena Corporation
Derma-Life Corporation
Desert Essence
Devi, Inc.
Dial Corporation
Dr. Hauschka Cosmetics
Earth Science, Inc.
Ecco Bella Botanicals
Ellon USA
Elysee Scientific Cosmetics
Espree Associates, Inc.
Estee Lauder Companies
Essque Bodycare
Eva Jon Cosmetics
Ever Young, Inc.
Faith Products, Ltd.
Flame Glow
Finelle Cosmetics
Framesi USA/Roffler
French Transit, Ltd.
Garden Botanika
Georgette Klinger, Inc.
Green Mountain
    Products, Inc.
Gryphon Development
Hawaiian Resources
    Co., Ltd.
Health From The Sun
Helen Lee Skin Care & Cos.
Heritage Store, Inc.
Homebody/Perfumoils, Inc.
I-Tech Laboratories
International
Izy's Skin Care Products
J & J Jojoba/California
    Gold Products
Jafra Cosmetics, Inc.
Janca's Jojoba Oil &
    Seed Company
Janet Sartin Cosmetics
Jessica McClintock

Joe Blasco Cosmetics
Jolen, Inc.
Juvenesse
Katonah Scentral
Keren Happuch, Ltd.
Kiss My Face Corp.
Klaire Laboratories, Inc.
KSA Jojoba
L'Herbier De Provence, Ltd.
La Dove, Inc.
Lady Burd Exclusive Private
    Label Cosmetics
Lange Laboratories
Lily of Colorado
Logona USA, Inc.
Lotions & Potions
Lotus Brands
Lotus Light Enterprises
Luzier Personalized
    Cosmetics
Magic of Aloe, Inc.
Magick Botanicals/
    Magick Mud
Maharishi Ayur-Veda
    Products
Mar-Riche Enterprises, Inc.
Margarite Cosmetics/
    Moon Products, Inc.
Marilyn Miglin, L.P.
Mario Badescu
    Skincare, Inc.
Mary Kay Cosmetics, Inc.
Mastey De Paris, Inc.
Matrix Essentials, Inc.
Michel Constantini Cosme
Mountain Rose Herbs
Naturade
Nature's Acres
Nature's Colors Cosmetics
Nature's Elements
    International, Ltd.
Nature's Sunshine
    Products, Inc.
NaturElle Cosmetics Corp.
Nectar USA
New Moon Extracts
Neways, Inc.
OPI Products, Inc.
Oxyfresh USA, Inc.

Palm Beach Beauty Products
Patricia Allison
Penn Herb Company, Ltd.
Pepper's
Pharmagel
Philip B., Inc.
Phybiosis
Prestige Cosmetics
Pro-Ma Systems (USA), Inc.
Rachel Perry, Inc.
Reviva Labs, Inc.
Revlon, Inc.
Royal Labs Natural
    Cosmetics
Set-N-Me-Free
Shaklee U.S., Inc.
Smith & Vandiver, Inc.
Sombra Cosmetics, Inc.
Steps In Health, Ltd.
Studio Magic, Inc.
Sunrise Lane Products, Inc.
TAUT by Leonard Engelman
TCCD International, Inc.
TN Dickinson
Tisserand Aromatherapy
Ultra Glow Cosmetics
Vanda Beauty Counselor
Vegan Market
Venus & Apollo

Dental Care
Action Labs
Amberwood
Auromere Ayurvedic
    Imports
Basically Natural
Baudelaire, Inc.
Body Tools
Caring Catalog
Caswell-Massey
Classic Cosmetics, Inc.
Comfort Manufacturing
    Company
Compassion Matters
Dep Corporation
Desert Essence
Eco-Dent International, Inc.
EM Enterprises
Eva Jon Cosmetics

Health From The Sun
Heritage Store, Inc.
Internatural
Levlad, Inc.
Logona USA, Inc.
Lotions & Potions
Lotus Brands
Lotus Light Enterprises
Mountain Rose Herbs
Neways, Inc.
Nutribiotic
Peelu
Penn Herb Company, Ltd.
Precious Collection
Set-N-Me-Free
Shaklee U.S., Inc.
Steps In Health, Ltd.
Sunrise Lane Products, Inc.
Terressentials
Thursday Plantation
   Labs, Pty., Ltd.
Tom's of Maine
Universal Light
Vegan Market

**Deodorants/Antiper-
spirants**
Aloeprime, Inc.
Amberwood
American International Ind.
Aubrey Organics
Black Pearl Gardens
Body & Soul of Chicago
Body Shop, Inc.
Body Tools
Borlind of Germany
Canada's All Natural
   Soap, Inc.
Caring Catalog
Caswell-Massey
Christopher Enterprises
Coastline Products
Common Scents
Deodorant Stones
   of America (DSA)
Devi, Inc.
Dial Corporation
E. Burnham Cosmetic
   Company, Inc.

French Transit, Ltd.
Gryphon Development
Heritage Store
IQ Products Company
Izy's Skin Care Products
Kiss My Face Corp.
Logona USA, Inc.
Lotions & Potions
Louise Bianco Skin Care, Inc.
Luzier Personalized
   Cosmetics
Marilyn Miglin, L.P.
Medicine Flower, Inc.
Montagne Jeunesse
Natural Bodycare
Nature's Elements
   International, Ltd.
Nectar USA
Neways, Inc.
Palm Beach Beauty
   Products
Penn Herb Company, Ltd.
Perfect Natural Products
PlantEssence
Potions & Lotions/
   Body & Soul
Seventh Generation
Shaklee U.S., Inc.
Starwest Botanicals, Inc.
Steps In Health, Ltd.
Sunrise Lane
TCCD International, Inc.
Terra Flora Herbal
   Body Care Products
Terressentials
Third Millennium Science
Thursday Plantation, Inc.
Tom's of Maine
U.S. Sales Service
Vanda Beauty Counselor
Vegan Market
Wysong Corporation

Feminine Hygiene
Aloe Flex Products, Inc.
BioFilm, Inc.
Chanty, Inc.
Ever Young, Inc.
Glad Rags

Health From The Sun
Internatural
Les Femmes, Inc.
Lotus Brands
Lotus Pads
Mountain Rose Herbs
Natracare
Naturade
Precious Collection
Seventh Generation
Starwest Botanicals, Inc.
Thursday Plantation, Inc.

First Aid
Aloe Creme Laboratories
Aloe Flex Products, Inc.
Aloeprime, Inc.
Amberwood
Aubrey Organics
Botan Corporation
Breezy Balms
Carma Labs, Inc.
Chanty, Inc.
Christopher Enterprises
Derma-Life Corporation
Desert Essence
Green Mountain
   Products, Inc.
Herbal Products &
   Development
Heritage Store, Inc.
Joshua Solution
Leydet Aromatics
Mentholatum Company
   of Canada, Ltd.
Mountain Rose Herbs
Naturade
Nature's Acres
Nature's Pharmacy
Nutribiotic
Palm Beach Beauty Products
Penn Herb Company, Ltd.
Precious Collection
Set-N-Me-Free
Simplers Botanical Co.
Song of Life, Inc.
Starwest Botanicals, Inc.
Sunrise Lane Products, Inc.
Tender Corporation

Thursday Plantation, Inc.
Zenith Advanced Health
  Systems International

Foot Care
Action Labs
Aloe Flex Products, Inc.
Aloeprime, Inc.
American International Ind.
Aphrodisia Products, Inc.
Bare Escentuals
Bath Island, Inc.
Body & Soul of Chicago
Body Shop, Inc.
Body Time
Body Tools
Bodyography
Caswell-Massey
Classic Cosmetics, Inc.
Cleopatra's Secret
  from the Dead Sea
Concept Now Cosmetics
Creation Soaps, Inc.
Desert Essence
Dr. Hauschka Cosmetics
E. Burnham Cosmetic
  Company, Inc.
Earth Science, Inc.
Eco Pak Canada
Eva Jon Cosmetics
Exotic Nature
Focus 21International, Inc.
Garcoa Labs
Garden Botanika
Gena Labs
Genki USA
Goodier, Inc.
Gryphon Development
Health From The Sun
Helen Lee Skin Care & Cos.
Heritage Store, Inc.
Internatural
Janca's Jojoba Oil &
  Seed Company
Khepra Skin Care, Inc.
KSA Jojoba
Les Femmes, Inc.
Levlad, Inc.
Little Herbal Garden

Lotions & Potions
Lotus Brands
Luzier Personalized
  Cosmetics
Masada Marketing Company
Mountain Rose Herbs
Naturade
Natural Bodycare
Natural Way Natural
  Body Care
Nature's Corner
Nature's Elements
  International, Ltd.
Nectar USA
Nutribiotic
Palm Beach Beauty
  Products
Pepper's
Perfect Natural Products
Philip B., Inc.
Precious Collection
Sea Minerals Co.
Set-N-Me-Free
Shaklee U.S., Inc.
Smith & Vandiver, Inc.
Soap Factory
Strong Skin Savvy, Inc.
TCCD International, Inc.
Third Millennium Science
Thursday Plantation
  Labs, Pty., Ltd.
Tisserand Aromatherapy
U.S. Sales Service
Vanda Beauty Counselor

Fragrances
AKA Saunders, Inc.
Alexandra Avery Purely
  Natural Body Care
Amberwood
Ananda Country
  Products
Aroma Life Company
Aromaland, Inc.
Aubrey Organics
Auromere Ayurvedic
  Imports
Aveda Corporation
Bare Escentuals

Bath Island, Inc.
BeautiControl
  Cosmetics, Inc.
Bella's Secret Garden
Benetton Cosmetics Corp.
Black Pearl Gardens
Body & Soul of Chicago
Body Love Natural
  Cosmetics, Inc.
Body Shop, Inc.
Body Time
Body Tools
Bodyography
Bonne Bell, Inc.
Canada's All Natural
  Soap, Inc.
Caring Catalog
Cassini Parfums, Ltd.
Caswell-Massey
Chanel, Inc.
Chanty, Inc.
Chishti Company
Christian Dior Perfumes, Inc.
Clarins of Paris
Cloudworks
Common Scents
Compassion Matters
Creation Soaps, Inc.
Devi, Inc.
Ecco Bella Botanicals
Eden Botanicals
Elysee Scientific Cosmetics
Essential Aromatics
Finelle Cosmetics
Forest Essentials
Garden Botanika
Goodier, Inc.
Gryphon Development
Hawaiian Resources
  Co., Ltd.
Helen Lee Skin Care & Cos.
Herb Garden
Heritage Store, Inc.
Homebody/Perfumoils, Inc.
Infinite Quality, Inc.
Internatural
IQ Products Company
Janca's Jojoba Oil &
  Seed Company

Kama Sutra Company
Katonah Scentral
KSA Jojoba
L'Herbier De Provence, Ltd.
La Natura
Lady of the Lake Company
Leydet Aromatics
Little Herbal Garden
Lotions & Potions
Lotus Brands
Lotus Light Enterprises
Luzier Personalized
    Cosmetics
Marilyn Miglin, L.P.
Medicine Flower, Inc.
Montagne Jeunesse
Mountain Rose Herbs
Native Scents, Inc.
Natural Way Natural
    Body Care
Nature's Acres
Nature's Corner
Nature's Elements
    International, Ltd.
Nectar USA
Patricia Allison
Paul Mazzotta, Inc.
Pepper's
PlantEssence
Potions & Lotions/
    Body & Soul
Precious Collection
Royal Labs Natural
    Cosmetics
Shaklee U.S., Inc.
Simple Wisdom, Inc.
Simplers Botanical Co.
Smith & Vandiver, Inc.
Starwest Botanicals, Inc.
Sumeru Garden Herbals
Surco Products, Inc.
TerraNova
Terressentials
Third Millennium Science
Tisserand Aromatherapy
Uncommon Scents, Inc.
Vanda Beauty Counselor
Whole Spectrum
Windrose Trading Company

**Furniture Polishes,
Waxes, Cleaners**
AFM Enterprises
Allens Naturally
Amazon Premium Products
Austin's
Card Corporation
CHIP Distribution
    Company
Dial Corporation
Earthly Matters
Eco Pak Canada
EM Enterprises
Frank T. Ross & Sons, Ltd.
Golden Pride - Rawleigh
Huish Detergents, Inc.
IQ Products Company
Janca's Jojoba Oil &
    Seed Company
Jelmar Company
Lan-O-Sheen, Inc.
Parker & Bailey
Pets 'N' People, Inc.
RCN Products, Inc.
Seventh Generation
Third Millennium Science
Vegan Market
Zenith Advanced Health
    Systems International

Hair Care
ABBA Products, Inc.
Action Labs
Advanced Research Labs
AFM Enterprises
AKA Saunders, Inc.
Aloe Creme Laboratories
Aloe Flex Products, Inc.
Aloegen Natural Cosmetics
Amberwood
American International Ind.
Aphrodisia Products, Inc.
Aroma Life Company
Aromaland, Inc.
Aubrey Organics
Avalon Natural
    Cosmetics, Inc.
Aveda Corporation
Bare Escentuals

Basic Elements Hair
    Care Systems, Inc.
Basically Natural
Bath Island, Inc.
Beauty Naturally
Beauty Without
    Cruelty Cosmetics
BeautyMasters
Beehive Botanicals, Inc.
Bella's Secret Garden
Bio-Pac
Bio-Tec Cosmetics Ltd.
Black Pearl Gardens
Blessed Herbs
Body & Soul of Chicago
Body Shop, Inc.
Body Tools
Bodyography
Bonne Bell, Inc.
Borlind of Germany
Botan Corporation
Canada's All Natural
    Soap, Inc.
Caring Catalog
Carme, Inc.
Caswell-Massey
Cernitin America Inc.
Chanty, Inc.
Chempoint Products Inc.
Chenti Products, Inc.
Chica Bella, Inc.
Classic Cosmetics, Inc.
Cleopatra's Secret
    from the Dead Sea
Coastline Products
Colonial Dames Co. Ltd.
Colora
Compassion Matters
Conair Corporation
Concept Now Cosmetics
Creation Soaps, Inc.
Del Laboratories
Dena Corporation
Dep Corporation
Derma-Life Corporation
Desert Essence
Devi, Inc.
Dial Corporation
Dr. Hauschka Cosmetics

E. Burnham Cosmetic
 Company, Inc.
EM Enterprises
Earth Friendly Products
Earth Preserv, Ltd.
Earth Science, Inc.
Ecco Bella Botanicals
EM Enterprises
Essential Aromatics
Eva Jon Cosmetics
Faith Products, Ltd.
Finelle Cosmetics
Focus 21International, Inc.
Forest Essentials
Framesi USA/Roffler
Frank T. Ross & Sons, Ltd.
French Transit, Ltd.
Garcoa Labs
Garden Botanika
Gena Labs
Genki USA
Goodier, Inc.
Granny's International, Inc.
Green Mountain
 Products, Inc.
Greenway Products, Inc.
Gryphon Development
Hair Doc Company
Hawaiian Resources Co., Ltd.
Head Shampoo, Inc./
 Pure & Basic Products
Healthy Times
Helen Lee Skin Care & Cos.
Heritage Store, Inc.
Hobe Labs, Inc.
Homebody/Perfumoils, Inc.
Image Labs, Inc.
Infinite Quality, Inc.
Institute of Trichology
International
IQ Products Company
J & J Jojoba/California
 Gold Products
Janca's Jojoba Oil & Seed Co.
Joe Blasco Cosmetics
John Paul Mitchell Systems
JOICO Labs, Inc.
JR Liggett Ltd.
Katonah Scentral

Kiss My Face Corp.
Klaire Laboratories, Inc.
KMS Research, Inc.
KSA Jojoba
La Dove, Inc.
Lan-O-Sheen, Inc.
Lange Laboratories
Lanza Research Int'l
Les Femmes, Inc.
Levlad, Inc.
Logona USA, Inc.
Lotions & Potions
Lotus Brands
Lotus Light Enterprises
Luzier Personalized
 Cosmetics
Magic of Aloe, Inc.
Magick Botanicals/
 Magick Mud
Maharishi Ayur-Veda
 Products
Mario Badescu
 Skincare, Inc.
Masada Marketing Company
Mastey De Paris, Inc.
Matrix Essentials, Inc.
McLaughlin, Inc.
MEN by Geoff Thompson
Mera Personal
 Care Products
Michel Constantini Cosme
Montagne Jeunesse
Mountain Rose Herbs
Naturade
Natural Bodycare
Nature's Elements
 International, Ltd.
Nature's Plus
Nectar USA
New Moon Extracts, Inc.
Neways, Inc.
Nexxus Products Company
Nirvana, Inc.
North Country Glycerin
 Soap and Candle
Palm Beach Beauty Products
Patricia Allison
Paul Mazzotta, Inc.
Penn Herb Company, Ltd.

Pepper's
Perfect Natural Products
Pharmagel
Philip B., Inc.
Potions & Lotions/
 Body & Soul
Precious Collection
Premier One Products
Pro-Ma Systems (USA),
Inc.
RJF, Inc./Oshadhi
Royal Labs
 Natural Cosmetics
San Francisco Soap Company
Scruples Professional
Sea Minerals Co.
Set-N-Me-Free
Seventh Generation
Shaklee U.S., Inc.
Simple Wisdom, Inc.
Smith & Vandiver, Inc.
Soap Factory
Soap Works
Starwest Botanicals, Inc.
Steps In Health, Ltd.
Sumeru Garden Herbals
Sunfeather Herbal
 Soap Company
Sunrise Lane Products, Inc.
Susan Lucci Hair Care
TerraNova
Terressentials
Third Millennium Science
Thursday Plantation
 Labs, Pty., Ltd.
Tisserand Aromatherapy
TN Dickinson
Tom's of Maine
Trans-India(Shikai)
Travel Mates America
Vanda Beauty Counselor
Vegan Market
Venus & Apollo
Wella Corporation
Whole Spectrum
WiseWays Herbals
Wysong Corporation

Hair Perms & Coloring
ABBA Products, Inc.
Amberwood
American International Ind.
Aveda Corporation
Beauty Naturally
Bio-Tec Cosmetics Ltd.
Body Shop, Inc.
Chanty, Inc.
Colora
Conair Corporation
Dena Corporation
Dep Corporation
Focus 21International, Inc.
Framesi USA/Roffler
Image Labs, Inc.
Internatural
John Paul Mitchell Systems
JOICO Labs, Inc.
Katonah Scentral
KMS Research, Inc.
La Dove, Inc.
Lange Laboratories
Lanza Research International
Lotions & Potions
Lotus Brands
Mastey De Paris, Inc.
Mountain Rose Herbs
Nexxus Products Company
Palm Beach Beauty Products
Paul Mazzotta, Inc.
Scruples Professional
Salon Products
Starwest Botanicals, Inc.
Sunrise Lane Products, Inc.
Wella Corporation

Insect Repellents
Amberwood
American Eco-Systems
Aromaland, Inc.
Basically Natural
Body & Soul of Chicago
Body Tools
Breezy Balms
Caring Catalog
Christopher Enterprises
Color & Herbal Company
Copper-Brite, Inc.

Creation Soaps, Inc.
Don't Bug Me, Inc.
EM Enterprises
Essential Products
    of America, Inc.
Four (IV) Trail Products
Green Ban
Herb Garden
Herbal Products &
    Development
IQ Products Company
Janca's Jojoba Oil &
    Seed Company
Katonah Scentral
Lady of the Lake Company
Leydet Aromatics
Medicine Flower, Inc.
Mountain Rose Herbs
Natural Animal, Inc.
Nature's Elements
    International, Ltd.
New Methods
North Country Glycerin
    Soap and Candle
Paul Mazzotta, Inc.
Perfect Natural Products
Simple Wisdom, Inc.
Starwest Botanicals, Inc.
Sunfeather Herbal
    Soap Company
Tender Corporation
Third Millennium Science
Thursday Plantation
    Labs, Pty., Ltd.
Whole Spectrum

Laundry Products
Allens Naturally
Aloeprime, Inc.
Amberwood
America's Finest
    Products Corp.
American Eco-Systems
Amway Corporation
Austin's
Aveda Corporation
Basically Natural
Bi-O-Kleen Industries Inc.
Bio-Pac

Biogime
Caring Catalog
Chempoint Products, Inc.
CHIP Distribution Company
Coastline Products
Compassion Matters
Cot 'N' Wash
Country Save Corporation
Dial Corporation
Earth Friendly Products
Earthly Matters
Ecover, Inc.
EM Enterprises
Faith Products, Ltd.
Forever New Int'l, Inc.
Frank T. Ross & Sons, Ltd.
Golden Pride - Rawleigh
Granny's International, Inc.
Green Mountain
    Products, Inc.
Greenway Products, Inc.
Heritage Store, Inc.
Home Service Products
    Company
Huish Detergents, Inc.
Lan-O-Sheen, Inc.
Lifeline Company
McLaughlin, Inc.
Mountain Rose Herbs
Natural Bodycare
Oasis Biocompatible
RCN Products, Inc.
Seventh Generation
Shaklee U.S., Inc.
Sierra Dawn Products
Simple Wisdom, Inc.
Soap Factory
Sunrise Lane Products, Inc.
Val-Chem Company Inc.
Vegan Market
Zenith Advanced Health
    Systems International

Massage
AKA Saunders, Inc.
Alexandra Avery Purely
    Natural Body Care
Aloe Flex Products, Inc.
Amberwood

Ananda Country Products
Aphrodisia Products, Inc.
Aroma Life Company
Aromaland, Inc.
Auromere Ayurvedic
    Imports
Avalon Natural
    Cosmetics, Inc.
Aveda Corporation
Bare Escentuals
Bavarian Alpenol &
    Sunspirit
Bella's Secret Garden
Black Pearl Gardens
Body & Soul of Chicago
Body Love Natural
    Cosmetics, Inc.
Body Shop, Inc.
Body Time
Body Tools
Bodyography
Borlind of Germany
California Olive Oil Corp.
Caribbean Pacific of
    the Rockies
Caswell-Massey
Classic Cosmetics, Inc.
Cloudworks
Come To Your Senses
Common Scents
Compassion Matters
Creation Soaps, Inc.
Creme de la Terre
Desert Essence
Devi, Inc.
Dr. Hauschka Cosmetics
Dry Creek Herb Farm
Earth Science, Inc.
Ecco Bella Botanicals
Essential Aromatics
Essential Elements
Essential Products of
    America, Inc.
Exotic Nature
Faith Products, Ltd.
Forest Essentials
Free Spirit Enterprises
Garden Botanika
Gena Labs

Goodier, Inc.
Granny's International, Inc.
Green Ban
Green Mountain
    Products, Inc.
Gryphon Development
Hawaiian Resources Co., Ltd.
Healthy Times
Helen Lee Skin Care & Cos.
Herb Garden
Heritage Store, Inc.
Homebody/Perfumoils, Inc.
Internatural
J & J Jojoba/California
    Gold Products
Janca's Jojoba Oil &
    Seed Company
Kama Sutra Company
Katonah Scentral
Khepra Skin Care, Inc.
Kiss My Face Corp.
KSA Jojoba
La Dove, Inc.
La Natura
Lady of the Lake Company
Les Femmes, Inc.
Leydet Aromatics
Little Herbal Garden
Lotions & Potions
Lotus Brands
Lotus Light Enterprises
Lunar Farms
    Herbal Specialist
Maharishi Ayur-Veda
    Products
Mar-Riche Enterprises, Inc.
Medicine Flower, Inc.
Michael's Naturopathic
    Programs
Michel Constantini Cosme
Montagne Jeunesse
Mountain Rose Herbs
Murad, Inc.
Nadina's Cremes
Naturade
Natural Bodycare
Natural Way Natural
    Body Care
Nature's Acres

Nature's Corner
Nature's Elements Int'l, Ltd.
Nectar USA
Neways, Inc.
Norimoor Company, Inc.
Palm Beach Beauty Products
Patricia Allison
Paul Mazzotta, Inc.
Penn Herb Company, Ltd.
Pepper's
Phybiosis
PlantEssence
Potions & Lotions/
    Body & Soul
Precious Collection
Quan Yin Essentials
Ravenwood
RJF, Inc./Oshadhi
Royal Labs Natural
    Cosmetics
Set-N-Me-Free
Shirley Price
    Aromatherapy, Ltd.
Simple Wisdom, Inc.
Simplers Botanical Co.
Sinclair & Valentine
Smith & Vandiver, Inc.
Starwest Botanicals, Inc.
Sumeru Garden Herbals
TerraNova
Terressentials
Third Millennium Science
Thursday Plantation, Inc.
Tisserand Aromatherapy
U.S. Sales Service
Uncommon Scents, Inc.
Whole Spectrum
WiseWays Herbals

Nail Care
Action Labs
Amberwood
American International Ind.
Aphrodisia Products, Inc.
Avalon Natural
    Cosmetics, Inc.
Bare Escentuals
Barristo, Ltd.

BeautiControl
Cosmetics, Inc.
Beauty Without
Cruelty Cosmetics
BeautyMasters
Blue Cross Beauty
Products, Inc.
Borlind of Germany
Botan Corporation
Caring Catalog
Caswell-Massey
Clarins of Paris
Color My Image, Inc.
Compassion Matters
Del Laboratories
Dr. Hauschka Cosmetics
Earth Science, Inc.
Focus 21International, Inc.
Garden Botanika
Gena Labs
Genki USA
Health From The Sun
Helen Lee Skin Care & Cos.
Internatural
IQ Products Company
Janca's Jojoba Oil &
Seed Company
Keren Happuch, Ltd.
Khepra Skin Care, Inc.
KSA Jojoba
La Dove, Inc.
Lange Laboratories
Leydet Aromatics
Lotus Brands
Luzier Personalized
Cosmetics
Magic of Aloe, Inc.
Naturade
Natural Way Natural
Body Care
Nature's Elements
International, Ltd.
Nature's Plus
NaturElle Cosmetics Corp.
Nectar USA
Neways, Inc.
OPI Products, Inc.
Palm Beach Beauty Products
Paul Mazzotta, Inc.

Pharmagel
Philip B., Inc.
Prestige Cosmetics
Pro-Ma Systems (USA), Inc.
Set-N-Me-Free
Thursday Plantation, Inc.
Tisserand Aromatherapy
U.S. Sales Service
Vanda Beauty Counselor
Venus & Apollo

Paper Products
Earth Care
Earthly Matters
Fort Howard Corporation
Herb Garden
Lotions & Potions
Seventh Generation
Wirth International

Plant Care
C.E. Jamieson & Co., Ltd.
Colora
Compassion Matters
Ellon USA
Janca's Jojoba Oil &
Seed Company
McLAughlin, Inc.
Shaklee U.S., Inc.
St. Clair Industries, Inc.

**Potpourri/Sachets/
Incense**
Aphrodisia Products, Inc.
Aveda Corporation
Bath Island, Inc.
Bella's Secret Garden
Black Pearl Gardens
Body & Soul of Chicago
Body Encounters
Body Shop, Inc.
Body Time
Body Tools
Caswell-Massey
Common Scents
Essential Aromatics
Forest Essentials
Forever New Int'l, Inc.
Garden Botanika

Gryphon Development
Helen Lee Skin Care & Cos.
Herb Garden
Heritage Store, Inc.
Internatural
Katonah Scentral
L'Herbier De Provence, Ltd.
Little Herbal Garden
Lotions & Potions
Lotus Brands
Lotus Light Enterprises
Mountain Rose Herbs
Native Scents, Inc.
Natural Bodycare
Nature's Corner
Nature's Elements
International, Ltd.
Nectar USA
Paul Mazzotta, Inc.
Penn Herb Company, Ltd.
Potions & Lotions/
Body & Soul
Ravenwood
Royal Labs Natural
Cosmetics
Sarah Michaels
Smith & Vandiver, Inc.
Starwest Botanicals, Inc.
Terressentials
Windrose Trading Company
WiseWays Herbals

Shaving Products
Alba Naturals, Inc.
Alexandra Avery Purely
Natural Body Care
Amberwood
American International Ind.
Aubrey Organics
Aveda Corporation
Bare Escentuals
Basically Natural
Bath Island, Inc.
Body & Soul of Chicago
Body Shop, Inc.
Body Time
Botan Corporation
Brookside Soap Company
Brush Craft

Canada's All Natural
   Soap, Inc.
Caring Catalog
Caswell-Massey
Coastline Products
Comfort Manufacturing
   Company
Common Scents
Compassion Matters
Creation Soaps, Inc.
Derma-E Body Care
Desert Essence
Earth Science, Inc.
EM Enterprises
Focus 21International, Inc.
Garden Botanika
Goodier, Inc.
Gryphon Development
International
Izy's Skin Care Products
Janca's Jojoba Oil &
   Seed Company
Katonah Scentral
Kiss My Face Corp.
Les Femmes, Inc.
Logona USA, Inc.
Lotions & Potions
Lotus Brands
Maharishi Ayur-Veda
   Products
Mario Badescu Skincare, Inc.
MEN by Geoff Thompson
Natural Way Natural
   Body Care
Nature's Elements
   International, Ltd.
Nectar USA
Neways, Inc.
Nexxus Products Company
Pepper's
Perfect Natural Products
Potions & Lotions/
   Body & Soul
Royal Labs Natural
   Cosmetics
Seventh Generation
Smith & Vandiver, Inc.
Stature Field Corporation
Sunrise Lane Products, Inc.

Terressentials
Tom's of Maine
Vanda Beauty Counselor
Vegan Market
Wilkinson Sword, Inc.
Wysong Corporation
Zenith Advanced Health
   Systems International

Shoe Polish
Eco Pak Canada
Janca's Jojoba Oil &
   Seed Company

Skin Care
ABBA Products, Inc.
Action Labs
AKA Saunders, Inc.
Alba Naturals, Inc.
Alexandra Avery Purely
   Natural Body Care
Alivio Products, Inc.
Aloe Creme Laboratories
Aloe Flex Products, Inc.
Aloegen Natural Cosmetics
Aloeprime, Inc.
Amberwood
American International Ind.
American ORSA, Inc.
Ancient Formulas, Inc.
Andrew Jergens Company
Aphrodisia Products, Inc.
Aroma Life Company
Aromaland, Inc.
Aubrey Organics
Auromere Ayurvedic
   Imports
Autumn Harp, Inc.
Avalon Natural
   Cosmetics, Inc.
Aveda Corporation
Bare Escentuals
Basic Elements Hair
   Care System, Inc.
Basically Natural
Bath Island, Inc.
BeautiControl
   Cosmetics,Inc.
Beauty Naturally

Beauty Without
   Cruelty Cosmetics
BeautyMasters
Beehive Botanicals, Inc.
Beirsdorf, Inc.
Bella's Secret Garden
Biogime
Black Pearl Gardens
Body & Soul of Chicago
Body Encounters
Body Love Natural
   Cosmetics, Inc.
Body Shop, Inc.
Body Time
Body Tools
Bodyography
Bonne Bell, Inc.
Borlind of Germany
Botan Corporation
Botanics of California
Botanics Skin Care
Brookside Soap Company
C.E. Jamieson & Co., Ltd.
California Olive Oil Corp.
Canada's All Natural
   Soap, Inc.
Caribbean Pacific of
   the Rockies
Carma Labs, Inc.
Caring Catalog
Carme, Inc.
Caswell-Massey
Cernitin America, Inc.
Chanty Inc.
Chempoint Products, Inc.
Chenti Products, Inc.
Chica Bella, Inc.
Christopher Enterprises
Citra Brands, Inc.
Clarins of Paris
Classic Cosmetics, Inc.
Cloudworks
Coastline Products
Colin Ingram
Colonial Dames Co., Ltd.
Color Me Beautiful
Color My Image, Inc.
Compassion Matters
Conair Corporation

Concept Now Cosmetics
Country Comfort
Creation Soaps, Inc.
Creme de la Terre
Del Laboratories
Dep Corporation
Derma-E Body Care
Derma-Life Corporation
Dermatone Lab, Inc.
Desert Essence
Desert Naturels
Devi, Inc.
Dial Corporation
Dr. Hauschka Cosmetics
Dry Creek Herb Farm
E. Burnham Cosmetic
     Company, Inc.
Earth Preserve, Ltd.
Earth Science, Inc.
Ecco Bella Botanicals
Ella Bache, Inc.
Elysee Scientific Cosmetics
EM Enterprises
Espree Associates, Inc.
Essential Aromatics
Essential Products of
     America
Estee Lauder Companies
European Gold
European Soaps, Ltd.
Eva Jon Cosmetics
Ever Young, Inc.
Exotic Nature
Faith Products, Ltd.
Finelle Cosmetics
Focus 21International, Inc.
Forest Essentials
Free Spirit Enterprises
Garcoa Labs
Garden Botanika
Gena Labs
Genki USA
Georgette Klinger, Inc.
Goodier, Inc.
Great Mother's Goods
Green Mountain Products
Greenway Products, Inc.
Gryphon Development
Hawaiian Resources Co., Ltd.

Head Shampoo, Inc./
     Pure & Basic Products
Health From The Sun
Helen Lee Skin Care & Cos.
Heritage Store, Inc.
Hobe Labs, Inc.
Homebody/Perfumoils, Inc.
Infinite Quality, Inc.
Institute of Trichology
Internatural
IQ Products Company
Izy's Skin Care Products
J & J Jojoba/California
     Gold Products
Janca's Jojoba Oil &
     Seed Company
Janet Sartin Cosmetics
Joe Blasco Cosmetics
John Paul Mitchell Systems
Kama Sutra Company
Katonah Scentral
Keren Happuch, Ltd.
Khepra Skin Care, Inc.
Kiss My Face Corp.
KMS Research, Inc.
KSA Jojoba
La Dove, Inc.
La Natura
Lady Burd Exclusive
     Private Label Cosmetics
Lady of the Lake Company
Lange Laboratories
Les Femmes, Inc.
Levlad, Inc.
Leydet Aromatics
Lily of Colorado
Little Herbal Garden
Logona USA, Inc.
Lotions & Potions
Lotus Brands
Lotus Light Enterprises
Louise Bianco Skin Care, Inc.
Lunar Farms Herbal
     Specialist
Luzier Personalized
     Cosmetics
Magic of Aloe, Inc.
Magick Botanicals/
     Magick Mud

Maharishi Ayur-Veda
     Products
Mar-Riche Enterprises, Inc.
Marcha Labs, Inc.
Margarite Cosmetics/
     Moon Products, Inc.
Marilyn Miglin, L.P.
Mario Badescu Skincare, Inc.
Mastey De Paris, Inc.
Mary Kay Cosmetics
McLAughlin, Inc.
MEN by Geoff Thompson
Mera Personal Care
     Products
Michael's Naturopathic
     Programs
Michel Constantini Cosme
Montagne Jeunesse
Mother Love
     Herbal Company
Mountain Rose Herbs
Murad, Inc.
Nadina's Cremes
Naturade
Natural Attitudes, Inc.
Natural Bodycare
Natural Way Natural
     Body Care
Nature's Acres
Nature's Colors Cosmetics
Nature's Corner
Nature's Elements
     International, Ltd.
Nature's Plus
NaturElle Cosmetics Corp.
Nectar USA
New Moon Extracts, Inc.
Neways, Inc.
Nexxus Products Company
Nivea
Norimoor Company, Inc.
North Country Glycerin
     Soap and Candle
Nutri-Cell, Inc.
Nutribiotic
OPI Products, Inc.
Palm Beach Beauty
     Products
Patricia Allison

Paul Mazzotta, Inc.
Pepper's
Perfect Natural Products
Pharmagel
Philip B., Inc.
PlantEssence
Potions & Lotions/
    Body & Soul
Precious Collection
Pro-Ma Systems
    (USA), Inc.
Quan Yin Essentials
Rachel Perry, Inc.
Ravenwood
Reviva Labs, Inc.
Revlon, Inc.
RJF, Inc./Oshadhi
RJG Inc. (Formerly
    Dymer & Associates)
Royal Labs Natural
    Cosmetics
San Francisco Soap Company
Sarah Michaels
Schroeder & Tremayne Inc.
Sea Minerals Co.
Set-N-Me-Free
Seventh Generation
Shaklee U.S., Inc.
Shirley Price
    Aromatherapy Ltd.
Simple Wisdom, Inc.
Simplers Botanical Co.
Sinclair & Valentine
Smith & Vandiver Inc.
Soap Factory
Soap Works
Sombra Cosmetics Inc.
Song of Life, Inc.
Spirit of Saint Alban
St. Clair Industries, Inc.
Starwest Botanicals, Inc.
Stature Field Corporation
Steps In Health, Ltd.
Stick With Us Products
Strong Skin Savvy, Inc.
Sunrise Lane Products, Inc.
TAUT by Leonard Engelman
Terra Flora Herbal
    Body Care Products

TerraNova
Terressentials
Terry Laboratories, Inc.
Third Millennium Science
Thursday Plantation, Inc.
Tisserand Aromatherapy
Trans-India(Shikai)
Travel Mates America
Tropical Soap Company
U.S. Sales Service
Uncommon Scents, Inc.
Vanda Beauty Counselor
Vegan Market
Venus & Apollo
Vermont Country Soap
WiseWays Herbals
Wysong Corporation
Zenith Advanced Health
    Systems International

Suntan Care
Action Labs
Alba Naturals, Inc.
Alexandra Avery Purely
    Natural Body Care
Aloe Creme Laboratories
Aloe Flex Products, Inc.
Aloe Up, Inc.
Amberwood
Aubrey Organics
Autumn Harp, Inc.
Aveda Corporation
Bare Escentuals
Basically Natural
Bath Island, Inc.
BeautiControl
    Cosmetics, Inc.
Biogime
Body & Soul of Chicago
Body Encounters
Body Shop, Inc.
Body Time
Body Tools
Borlind of Germany
Botanics of California
Caribbean Pacific of
    the Rockies
Caring Catalog
Carme, Inc.

Chica Bella, Inc.
Clarins of Paris
Classic Cosmetics, Inc.
Colonial Dames Co. Ltd.
Color My Image, Inc.
Compassion Matters
Concept Now Cosmetics
Creation Soaps, Inc.
Creme de la Terre
Derma-Life Corporation
Dermatone Lab, Inc.
Desert Essence
Devi, Inc.
Don't Bug Me, Inc.
Earth Science, Inc.
Ella Bache, Inc.
Elysee Scientific Cosmetics
EM Enterprises
European Gold
Ever Young, Inc.
Faith Products, Ltd.
Finelle Cosmetics
Finley Pharmaceuticals
Focus 21International, Inc.
Garden Botanika
Goodier, Inc.
Great Mother's Goods
Green Mountain Products
Gryphon Development
Hawaiian Resources Co., Ltd.
Health From The Sun
Helen Lee Skin Care & Cos.
Heritage Store, Inc.
Internatural
J & J Jojoba/California
    Gold Products
Janca's Jojoba Oil &
    Seed Company
Janet Sartin Cosmetics
Katonah Scentral
KSA Jojoba
La Dove, Inc.
Lange Laboratories
Levlad, Inc.
Little Herbal Garden
Logona USA, Inc.
Lotions & Potions
Lotus Brands
Louise Bianco Skin Care, Inc.

Luzier Personalized
    Cosmetics
Magic of Aloe, Inc.
Mar-Riche Enterprises, Inc.
Marilyn Miglin, L.P.
Mario Badescu
Mastey De Paris, Inc.
Mera Personal Care
    Products
Michael's Naturopathic
    Programs
Michel Constantini Cosme
Murad, Inc.
Nadina's Cremes
Nature's Corner
Nature's Elements
    International, Ltd.
Nectar USA
Neways, Inc.
North Country Glycerin
    Soap and Candle
Palm Beach Beauty Products
Patricia Allison
Paul Mazzotta, Inc.

Pepper's
Pharmagel
Potions & Lotions/
    Body & Soul
Pro-Ma Systems (USA), Inc.
Reviva Labs, Inc.
Royal Labs Natural
    Cosmetics
Set-N-Me-Free
Shaklee U.S., Inc.
Simple Wisdom, Inc.
Smith & Vandiver, Inc.
Soap Factory
St. Clair Industries, Inc.
Thursday Plantation
    Labs, Pty., Ltd.
Vegan Market
Vanda Beauty Counselor
Whole Spectrum
Wysong Corporation

Miscellaneous
AFM Enterprises
(paints)

American Eco-Systems
(air quality filters)

BioFilm, Inc.
 (personal lubricant)

Carma Laboratories, Inc.
(lip balm for cold sores)

Desert Whale Jojoba Co., Inc.
(jojoba oil)

I-Tech Laboratories
 *(Op-tics cosmetic system
for eyeglass and lens users)*

Joshua Solution
(cold-canker sore solution)

Les Femmes, Inc.
*(natural sanding/buffing
disks for hair removal)*

PARENT COMPANY
SUBSIDIARIES, DIVISIONS & BRANDS

This section introduces the consumer to companies owning more than ten brands with product names not identifiable with their parent company. Therefore you will not find all companies, subsidiaries, divisions, and brands listed. Please keep in mind that companies and brands are being bought and sold continually. Not every subsidiary or division tests their products on animals. Please refer to the main section of the book for specific information concerning each listing.

If you have any unanswered questions about a particular brand and would like to see it listed in the next edition, please contact us via the questionnaire in the back of the book. We look forward to your assistance in keeping this publication the most comprehensive listing available to the public.

KEY TO SYMBOLS

♥ CRUELTY FREE: Does not test products or ingredients on animals

♡ Ingredients MAY be tested on animals

▼ Tests products or ingredients on animals

▼ Alberto-Culver Company
Subsidiaries/Divisions:
Sally Beauty Company; St. Ives
Brands:
Bold Hold; Consort Products for Men; FDS; Kleen Guard; Motions; Static Guard; Swiss Formula; TCB; Tresemme Products for Women; VO5

♥ BeautiControl Cosmetics, Inc.
Brands:
Becoming Color; Charade Cologne; Dose of Color; Face Feminizer; Flare Cologne; Lip Apeel; Microderm; Nailogics; Regeneration; Sensuous Shadows with Silkenspheres; Sheer Protection; Shine Tech; Skinlogics; Spectaculash; Success Cologne; Sunlogics; Unbelievable Blush

▼ Bristol-Myers Squibb Company
Subsidiaries/Divisions:
Calgon Vestal Laboratories; Clairol, Inc;, ConvaTec; Genetic Systems Corp.; Matrix Essentials; Mead Johnson Nutritional Group; Monarch Crown Corp; Surgitek; Westwood-Squibb Pharmaceuticals; Xomed Treace; Zimmer, Inc.
Brands:
Alpha Keri; Ban; Chase; Fisherman's Friend; Fostex; Graneodin; Keri; Pre Sun; Sea Breeze; Tempra; Tickle

♥ Carme, Inc.
Brands:
Allercreme; Biotene; Carme; Country Roads; DuBarry; Jojoba Farms; Loanda; Mild & Natural;

Mill Creek; MoisturEyes; Mountain Herbery; Silver Fox; Sleepy Hollow

▼ Clairol, Inc.
Brands:
Balsam Color; Basic White; Beautiful Collection; ColorHold; Complements; Condition; Final Net; Finale; Frost & Tip; Glints; Herbal Essences; Infusion 23 ; Jazzing; Kaleidocolors; Lasting Color; Loving Care; Men's Choice; Miss Clairol; Motif; Natural Instincts; Nice'n' Easy; Option; Quiet Touch; Second Nature; Shimmer Lights; Summer Blonde; Torrids; Ultress; Vitalis

▼ Clorox Company
Subsidiaries/Divisions:
Kingsford Products Company
Brands:
Brita; Clorox; Combat Insecticide; Control Cat Litter; Formula 409; Fresh Step Cat Litter; Liquid-Plumer Drain Opener; Litter Green Cat Box Filler; Pine-Sol Cleaner; Scoop Fresh Scoopable Litter; Soft Scrub; Tackle; Tilex; Twice As Fresh

▼ Colgate-Palmolive Company
Subsidiaries/Divisions:
Hills' Pet Products, Inc.; Murphy-Phoenix Company; Princess House; Softsoap Enterprises; Sterno, Inc.; Vipont Pharmaceuticals/Colgate Hoyt-Gel-Kam
Brands:
Ajax; Axion; Cashmere Bouquet; Cold Power; Dermassage; Fab; Fabuloso; Handiwipes; Hill's; Irish Spring; Javex; Kirkman Borax; Klorin; Kolynos; La Croix; Mennen; Mersene; Murphy's Oil Soap; Palmolive; Pouss' mousse; Protex; Sesame Street Children's Bath; Softsoap; Softwash; Suavitel; Super Suds; Vel Beauty; Village Bath

♥ Conair Corporation
Brands:
Magical Mane; Hair Management for Men; Jheri Redding; Vitamin Therapy; California Shine; Natural Care; Just Wonderful; Grande Finale; One-n-Only; Beverly Hills; Value Plus; Biogena; Ginza

♡ Cosmair, Inc.
Subsidiaries/Divisions:
L'Oreal of Paris; Lancome; Maybelline; Ralph Lauren Fragrance Division; Redken Labs
Brands:
Anais Anais; Biotherm; Cacheral; Drakkar Nor; Giorgio Armani; Gloria Vanderbilt; Guy Laroche; Ralph Lauren; L'Oreal; Lancome; Lauren; Paloma Picasso; Polo Crest; Safari

♥ Delby System
Brands:
Bufette; Elephant Ear; Gem; Icy Velvet; Lift-Off; Mediterranean; Movie Stars; Petites; Pumice Stone; Puppetdears; Sponge Drops; Starlet O'Hara; Travelette; Wedges

♥ Dep Corporation
Brands:
Agree; Cornucopia Valley; Cuticura; Habanita; Halsa; Jean-Louis Scherrer; Jordan ; LA

Looks; Lavoris; Lilt; Molinard; Nature's Family; Nino Cerutti; Pears; Porcelana; Topol

💜 Dial Corporation
Brands:

Borax; Breck Hair Care; Brillo Pads; Cameo Copper; Dial; Diaper Sweet; Dobie Pads; Dutch; Fels Naptha; Garden Bouquet; Hilex; Magic Sizing; Manpower Deodorant; Natural Soap; Parson's Ammonia; Pure & Natural; Purex Detergents; Sno-Bowl Toilet Bowl Cleaner; Sno Drops; Spirit Soap; Sta-Flo; Sta-Puf; Sweetheart; Tone Soap; Trend

▼ DowBrands, Inc.
Brands:

Apple Pectin; Fantastic; Glass Plus; Nucleic A; Perma Soft; Spray-n-Wash; Stain Stick; Style; Style Plus; Vitale; Yes

💜 Estee Lauder Companies
Subsidiaries/Divisions:

Aramis; Estee Lauder, Inc.; Clinique Labs; Origins Natural Resources; Prescriptives, Inc.

Brands:

Aramis; Aramis 900; Aramis Devin; JHL; Tuscanny; Beautiful; Cinnabar; Clinique; Knowing; Lauder for Men; Private Collection; Spellbound; White Linen

▼ Gillette Company
Subsidiaries/Divisions:

Braun; Jafra Cosmetics; Lustrasilk Corp. of America; Oral-B Labs; Stationery Products Group

Brands:

Aapri; Adorn; Atra; Bare Elegance;

Big Body; Blue Blades; Braun; Brush Plus Shaving System; Clear Gel; Cool Wave; Curl Free; Curve 'n Body; Custom Plus; Daisy; Deep Magic; Dippity-Do; Double Edge; Dry Idea; Dry Look; Earth Born; Face Saver; Flair; Foamy Shaving Cream; Foot Guard; For Oily Hair Only; Good News; Happy Face; Heads Up; Home Perms; Hot One Shave Creams; Image Body Spray; Just Whistle; Liquid Paper; Lustrasilk; Mink Difference; Oral B; Parker; Platinum-Plus Blades; Right Guard; Sensor; Silkience; Soft & Dri; Super Blue; Swivel; Tame; Toni; Trac II; Waterman; White Rain; Wild Rain

💜 Green Mountain Products, Inc.
Subsidiaries/Divisions:

Olde Tyme 1881 Company

Brands:

Aloe Gold; Baby Massage; Cool Wash; Dura Green; Glass Mate; Golden Lotus; Kleen; Kleer; Rainforest; Secretary's Secret; Soft n' Fresh; Super Pine; Vegelatum; Winter White

💛 Guerlain, Inc.
Brands:

Evolution; Guerlain Issima; Jardins de Bagatelle; L'Heure Bleue; Les Meteorites; Mitsouko; Odelys; Samsara; Shalimar; Terracotta

▼ Johnson & Johnson
Subsidiaries/Divisions:

McNeil Specialty Products; Neutrogena Corp.; Penaten G.m.b.H.; Personal Products Co.; RoC S.A.; Vistakon

Brands:

Act Flouride Rinse; Acuvue;

Band-Aid; Carefree Panty Shields; Clean & Clear; o.b. Tampons; One Touch; Purpose; Reach; Serenity Guards; Shower to Shower; Sport Strip; Stayfree; Sundown; Sure & Natural; Surefit; Surevue; Wondergrip

▼ Kiwi Brands Inc.
Brands:

Ambi; Behold; Cavalier Shoe Polish; Endust Meltonian; Miracle White; Nu-Life; Propert's Shoe Polish; Tana; Tintex Fabric Dye; Ty-D-Bowl; Wood Preen Floor Wax

♡ L'Oreal
Brands:

Anais Anais; Armani; Biotherm; Cacharel; Drakkar Noir; Giorgio Armani; Gloria Vanderbilt; Guy Laroche; Lancome; Lauren by Ralph Lauren; Paloma Picasso; Parfums Guy Laroche; Performing Preference; Polo by Ralph Lauren; Safari by Ralph Lauren; Studio Line; Vanderbilt by Gloria Vanderbilt

▼ Lever Brothers
Brands:

All Detergents; Breeze; Caress; Dove; Drive; Final Touch; Lever 2000; Lifebuoy; Rinso; Shield; Signal Mouthwash; Snuggle; Sunlight; Supreme au Creme; Surf; Wisk

♡ Mem Company, Inc.
Subsidiaries/Divisions:
Tom Fields, Ltd.
Brands:

Aqua de Selva; Clean & Natural; English Leather; Fathom; Heaven Scent; Love's; Love's

Baby Soft; Musk; Timberline; Tinkerbell; Victor of Milano

♥ Palm Beach Beauty Products
Brands:

Alpha-Frutein; Collagen Plus; Firm & Fill; Forever 29; Gly-Miracle; Hair Saver; Hair Therapy; Medical; Nice N' Thick; Nucleic Plus; Palm Beach; Placenta Plus; Protein Plus; Qualla; Quick Tan; Radiant Glo; Skin Saver; Stress Relief; Stress Therapy; Sun Defiance; Ultra-Fresh; Vita-Fusion; Winter Tan

▼ Procter & Gamble Company
Subsidiaries/Divisions:
Giorgio Beverly Hills; Max Factor & Company, Noxell Corp.; Pantene Corp.; Richardson-Vicks; Vidal Sassoon
Brands:

Always; Attends; Banner; Benzodent; Biz; Bold; Bounce; Bounty; Camay; Cascade; Charmin; Cheer; Clarion; Clearasil; Clearstick; Coast; Comet; Complete; Cover Girl; Crest; Dash; Dawn; Denquel Toothpaste; Downy; Dreft; Era; Fasteeth Denture Adhesive; Fixodent; Gain; Giorgio Beverly Hills; Gleem; Head & Shoulders; Hugo Boss; Incognito; Ivory; Joy Dishwashing Liquid; Kirk's; Kleenite; Laura Biagiotti-Roma; Lava; Le Jardin; Lestoil; Luvs; Mr. Clean; NaVy; Noxema; Oil of Olay; Old Spice; Oxydol; Pampers; Pantene; Pert; Prell; Puffs; Safeguard; Scope; Secret; Solo; Spic & Span; Summit; Sure;

Tide; Top Job; Toujours Moi;
Venezia; Vidal Sassoon; Zest

▼ **Reckitt & Colman, Inc.**
Subsidiaries/Divisions:
Airwick Industries; Boyle-
Midway Household Products
Brands:
Aerowax; Armstrong Floor
Cleaners; Black Flag; Brasso Metal
Cleaners; Bully Toilet Bowl
Cleaner; Carpet Fresh; Chore Boy;
Drain Power; Easy Wash; Easy-
Off; Glamorene Carpet Cleaner;
Lewis Red Devil Lye; Magic
Mushroom; Noxon Metal Polish;
Pam Non-Stick Cooking Spray;
Pan Handl'rs & Golden Fleece
Cleaning Pads; Quick Dip Silver
Cleaner; Rug Fresh; Sani-Flush;
Spray 'n Vac; Stick-Ups; Swish
Toilet Bowl Cleaner; Wizard Air
Freshener; Woolite; Zud Cleanser

❤ **Revlon, Inc.**
Subsidiaries/Divisions:
Almay, Inc.; Bill Blass, Inc.; Charles
Revson, Inc.; Charles of the Ritz
Group Ltd.; DVF, Inc.; Germain
Monteil Cosmetiques Corp.;
Halston Enterprises, Inc.; Lancaster,
Inc.; National Health Labs; Norell
Perfumes, Inc.; Princess Marcella
Borghese, Inc.; Roux Labs, Inc.;
Visage Beaute Cosmetics, Inc.
Brands:
Almay; Bill Blass; Borghese; Charles
of the Ritz; Chaz; Flex; Halston; Jean
Nate; Jontue; Lancaster; Norell;
Scoundrel; That Man; Ultima

▼ **S.C. Johnson & Son, Inc.**
Subsidiaries/Divisions:
Drackett Company; Johnson
Wax Development Corp.
Brands:
Armstrong Cleaner; Aveeno;
Befresh; Brite; Drackett Company;
Drano; Duster Plus; Edge Shave;
Envy; Favor; Future; Glade; Glo-
Coat; Glory Foam; Good Measure;
Good Sense; Johnson's Paste Wax;
Jubilee; Klean 'n Shine; Klear;
Liquid Glory; Mr. Muscle; Off!;
One Step; Pledge; Raid; Renuzit;
Shout; Step Saver; Twinkle; Vanish;
Wall Power; Windex; Woodrich

▼ **SmithKline Beecham
Consumer Brands**
Subsidiaries/Divisions:
SK & F Labs; Int'l Affiliated Labs
Brands:
A-200 Pyrinate; Aqua Fresh; Aqua
Velva; Brylcreem; Clear by Design;
Esoterica; Lectric Shave; Macleans;
Massengill; Orafix Denture Adhe-
sive; Oxy Products; Scott's Emulsion;
Vital Eyes; Williams Shave Products

▼ **Unilever United States, Inc.**
Subsidiaries/Divisions:
Calvin Klein Cosmetics Company;
Chesebrough-Ponds USA;
Elizabeth Arden Company; Lever
Brothers; National Starch &
Chemical; Prince Matchabelli;
Thomas J. Lipton Co.
Brands:
Baron; Burberry's for Men; Carol
Richards; Chloe; Cutex; Faberge;
Great Lady; KL Fragrance;
Lagerfeld; Lux; Most Precious;
Sea & Ski; Shadow Perfume; Tan
Accelerator; Tigress; White
Shoulders

CHARITIES WHICH DO NOT FUND ANIMAL RESEARCH

American Kidney Fund
6110 Executive Boulevard
Suite 1010
Rockville, MD 20852

Arthritis Research Institute of
America
300 S. Duncan Avenue, Suite 240
Clearwater, FL 34615

Association of Birth Defect
Children
827 Irma Avenue
Orlando, FL 32803

Cancer Care, Inc.
1180 Avenue of the Americas
New York, NY 10036

Cancer Fund of America, Inc.
2901 Breezewood Lane
Knoxville, TN 37921-1099

Designer Institute Foundation
for AIDS
150 West 26th Street, Suite 602
New York, NY 10001

Disabled American Veterans
P.O. Box 14301
Cincinnati, OH 45250-0301

Easter Seals
230 West Monroe Street
Suite 1800
Chicago, IL 60606-4703

The Green Foundation, Inc.
9481 Lechner Road
Fort Worth, TX 76179-4055

Heimlich Foundation
2368 Victory Parkway, Suite 410
Cincinnati, OH 45206

International Child Health Foundation
American City Building
P.O. Box 1205
Columbia, MD 21044

International Eye Foundation
7801 Norfolk Avenue
Bethesda, MD 20814

Multiple Sclerosis Association of
America
601 White Horse Pike
Oaklyn, NJ 08107

National Burn Victim Foundation
32-34 Scotland Road
Orange, NJ 07050

National Federation of the Blind
1800 Johnson Street, Suite 300
Baltimore, MD 21230-4998

National Head Injury Foundation
1776 Massachusetts Avenue, NW
Suite 100
Washington, DC 20036-1904

Quest Cancer Test
Woodbury, Harlow Road
Roydon, Harlow
Essex CM19 5HF
United Kingdom

The Rheumatoid Disease Foundation (aka The Arthritis Fund)
5106 Harding Road
Franklin, TN 37064

CHARITIES WHICH <u>STILL</u> FUND ANIMAL RESEARCH

Alzheimer's Disease and Related
Disorders Association
919 N. Michigan Avenue, Suite 1000
Chicago, IL 60611-1676

American Cancer Society
1599 Clifton Road, NE
Atlanta, GA 30329

American Diabetes Association
1660 Duke Street
Alexandria, VA 22314

American Heart Association
7320 Greenville Avenue
Dallas, TX 75231-4599

American Institute for Cancer Research
1759 R Street, NW
Washington, DC 20069

American Lung Association
National Headquarters
1740 Broadway
New York, NY 10019

American Parkinson Disease Assn.
60 Bay Street
Staten Island, NY 10301

Arthritis Foundation
1314 Spring Street NW
Atlanta, GA 30309

Cancer Prevention Project
1120 Connecticut Avenue, Suite 303
Washington, DC 20069

City of Hope
30 West 26th Street, Suite 301
New York, NY 10010

Cystic Fibrosis Foundation
6981 Arlington Road
Bethesda, MD 20814

Epilepsy Foundation of America
4351 Garden City Drive
Landover, MD 20785

The Foundation Fighting Blindness
(formerly National Retinitis
Pigmentosa Foundation)
1401 Mt. Royal Avenue
Baltimore, MD 21217

Joslin Diabetes Center
One Joslin Place
Boston, MA 02215

Leukemia Society of America
600 Third Avenue
New York, NY 10016

March of Dimes Birth Defects
Foundation
1275 Mamaroneck Avenue
White Palins, NY 10605

Muscular Dystrophy Association
3561 East Sunrise Drive
Tucson, AZ 85718

National Foundation for
Cancer Research
7315 Wisconsin Avenue, Suite 332W
Bethesda, MD 20814

National Kidney Foundation
30 E. 33rd Street
New York, NY 10016

National Multiple Sclerosis Society
733 Third Avenue
New York, NY 10017-3288

National Parkinson Foundation
1501 9th Avenue
Miami, FL 33136

National Psoriasis Foundation
660 SW 92nd Avenue
Portland, OR 97223-7195

Nina Hyde Center for Breast
Cancer Research
Lombardi Cancer Research Center
3800 Reservoir Road, NW
Washington, DC 20007

Parkinson's Disease Foundation
650 West 168th Street
New York, NY   10032-9982

Shriners Burn Institute
51 Blossom Street
Boston, MA 02114

St. Jude Children's Research
Hospital
National Executive Offices
1 St. Jude Place Bldg.
P.O. Box 3704
Memphis, TN 38173-0704

United Parkinson Foundation
360 West Superior Street
Chicago, IL 60610

University of Texas MD
Anderson Cancer Center
1515 Holcombe Boulevard, #135
Houston, TX 77030

Courtesy of Physicians Committee for Responsible Medicine, 5100 Wisconsin
Avenue, Suite 404, Washington, DC 20016.

Merchandise available through NAVS

T-shirts, Mugs, Pencils, Buttons

Books

Posters

Books	Price	Quantity	Amount
The Monkey Wars (Blum) **NEW!**	$15.00		
Declaration of the Rights of Animals	$10.00		
Personal Care for People Who Care	$4.95		
The Great Ape Project (Cavalieri & Singer)	$11.85		
Christianity & the Rights of Animals (Linzey)	$12.95		
Of Mice, Models, and Men (Rowan)	$19.95		
Animals in Education (Hepner)	$11.65		
Animals and Christianity (Linzey & Regan)	$14.95		
Diet for A New America (Robbins)	$13.95		
The Cruel Deception (Sharpe)	$12.00		
Animal Liberation (Singer)	$12.95		
Science, Animals & Evolution (Roberts)	$5.95		
67 Ways to Save the Animals (Sequoia)	$4.95		
Facing the Challenge (Cohen, Natelson)	$10.00		
Why Do Vegetarians Eat Like That (Gabbe)	$8.40		
Dog Scrap Book	$1.00		
NAVS Co-Sponsored Books			
Maternal Deprivation	$4.95		
Heart Research on Animals	$4.95		
Cancer Research	$4.95		
Merchandise			
Declaration of the Rights of Animals (sm. poster)	$3.00		
Declaration of the Rights of Animals (lrg. poster)	$7.00		
Art for Animals Classic - Poster: Still Life (cat)	$6.00		
Art for Animals Classic - Poster: Research (chimp)	$6.00		
Art for Animals Classic - Poster: Together (world)	$6.00		
Art for Animals Classic - Stickers (32)	$1.00		
World of Compassion Stickers (24)	$1.00		
Quotes of Compassion Stickers (32)	$1.00		
"Respect, Don't Dissect" Stickers - Frog (32)	$1.00		
"Laps Not Labs" Stickers - Cat (32)	$1.00		
"Live To Love and Let Live" Stickers (32)	$1.00		
"The Greatness of a Nation..." Bumper Stickers	$1.00		
"True Beauty is Living Cruelty Free" Button-Rabbit	$1.00		
"Respect Don't Dissect" Mug	$6.00		
Quotes of Compassion Mug	$6.00		
Personal Care Rabbit Pencil	$1.25		
World of Compassion Notecards (10 pack)	$5.00		
"Respect, Don't Dissect" T-shirt (L or XL only)	$15.00		
"Have A Heart..." T-shirt (L or XL only)	$15.00		
Global Awareness T-shirt (L or XL only)	$15.00		
It Doesn't Take A Genius Sweatshirt (M or L only)	$20.00		
Target NOAH Iron-on Logo	$2.00		
Target NOAH Bumper Sticker	$1.00		
Target NOAH Mug	$3.00		

Shipping Charges		
Up to $5.99 = $1.50 **VISA**	**Shipping Costs**	
$6.00 - $11.99 = $2.50		
$12.00 - $17.99 = $3.50 **MasterCard**	Total	
Each add'l $6.00 = $1.50		

**TO ORDER, CALL: 1-800-888-NAVS, or send this form and your
check or money order to: NAVS, Department 530-0110W,
P.O. Box 94020, Palatine, IL 60094-4020**

Notes

NAVS QUESTIONNAIRE

We'd like your help!

Please complete the following survey and return it to:
The National Anti-Vivisection Society
53 West Jackson Blvd., Ste. 1552, Chicago, IL 60604

This questionnaire is intended for research use only. The information contained herein will not be used for mailing lists or other promotional purposes.

Did you have any trouble finding companies or manufacturers?
_____ If so, explain

_____

_____

Is there any other information you would like to see included in future editions of this book? _____

_____

_____

How did you find out about this book? _____

_____

Where did you purchase this book? _____

_____

Are you a member of NAVS?   ☐ Yes ☐ No

Are you: ☐ Male ☐ Female

How old are you?
☐ 18 - 24   ☐ 25 - 34   ☐ 35 - 44   ☐ 45 - 54   ☐ 55+

What is your education level?
☐ High School   ☐ College   ☐ Advanced Degree

Questionnaire

After looking through this book, are there any companies or brands that did not appear that you would like to know about?

Name _____
Address _____
City/State/Zip _____
Phone _____

Name _____
Address _____
City/State/Zip _____
Phone _____

Name _____
Address _____
City/State/Zip _____
Phone _____

Name _____
Address _____
City/State/Zip _____
Phone _____

Name _____
Address _____
City/State/Zip _____
Phone _____

Thank you for helping us improve this book.

MEMBERSHIP APPLICATIONS

☐ Please send me **Personal Care for People Who Care** at a cost of $4.95 per book. Enclosed is my check for $ _____

☐ I would like to join NAVS. I understand that my membership includes one free copy of Personal Care.

Annual Memberships	**Lifetime Memberships**
☐ Individual $25	☐ Life Sponsor $100
☐ Student $10	☐ Life Benefactor $500
☐ Senior $12	☐ Life Partner $1000
☐ Other $ _____	☐ Please send me more information

Name _____

Address _____

City/State/Zip _____

Credit Card # _____ Expiration Date _____

☐ Visa ☐ MasterCard

Phone _____

Signature _____

Mail to: **The National Anti-Vivisection Society
Department 530-0110W
P.O. Box 94020
Palatine, IL 60094-4020**

Check, money order or credit card. No Cash. All contributions are tax-deductible to the fullest extent allowed by law.

Code #ABC07 VISA MasterCard To order today, call 1-800-888-NAVS (6287)

☐ Please send me **Personal Care for People Who Care** at a cost of $4.95 per book. Enclosed is my check for $ _____

☐ I would like to join NAVS. I understand that my membership includes one free copy of Personal Care.

Annual Memberships	**Lifetime Memberships**
☐ Individual $25	☐ Life Sponsor $100
☐ Student $10	☐ Life Benefactor $500
☐ Senior $12	☐ Life Partner $1000
☐ Other $ _____	☐ Please send me more information

Name _____

Address _____

City/State/Zip _____

Credit Card # _____ Expiration Date _____

☐ Visa ☐ MasterCard

Phone _____

Signature _____

Mail to: **The National Anti-Vivisection Society
Department 530-0110W
P.O. Box 94020
Palatine, IL 60094-4020**

Check, money order or credit card. No Cash. All contributions are tax-deductible to the fullest extent allowed by law.

Code #ABC08 VISA MasterCard To order today, call 1-800-888-NAVS (6287)

Notes

GIVE A GIFT OF COMPASSION
Personal Care for People Who Care

☐ Please send Personal Care for People Who Care at a cost of $4.95 per book. Enclosed is my check for $ _____

☐ I would like to give a gift of a NAVS membership. I understand the membership includes one free copy of Personal Care.

(Please check type of membership)

Annual Memberships
☐ Individual $25
☐ Student $10
☐ Senior $12

Lifetime Memberships
☐ Life Sponsor $100
☐ Life Benefactor $500
☐ Life Partner $1000

SEND TO:

Name: _____

Address: _____

City/State/Zip: _____

Phone: _____

FROM:

Name: _____

Address: _____

City/State/Zip: _____

Credit Card Acc't. # _____   Expiration Date _____

☐ Visa ☐ Master Card _____   Phone _____

Print name as it appears on the card _____

Signature _____

Check, money order or credit card. Please no cash.

All contributions are tax-deductible to the fullest extent allowed by law.

Mail to: **National Anti-Vivisection Society**
Department 530-0110W
P.O. Box 94020
Palatine, IL 60094-4020

To order today, call 1-800-888-NAVS

Code #ABC09   VISA   MasterCard

Notes